Praise for *The Story o*

Lloyd Alter is the most thoughtful, creative, funny, lucid, provocative person writing today about design. Probably nobody will agree with everything in this book. But every reader will learn much that is fresh and important.

— Denis Hayes, developer of the Bullitt Center,
and national coordinator, Earth Day

Lloyd Alter's *The Story of Upfront Carbon* is the book the world needs: fascinating, clear, and positive. Read this, get everyone you know to read it, and save the planet—as Alter shows, it's much easier to achieve than you may think.

— Dr. Barnabas Calder, Head of the History of Architecture
Research Cluster, University of Liverpool School of Architecture

To achieve a sustainable future, it's not enough to start using electric vehicles, energy-efficient appliances, and the like. It's a start, but the overwhelming majority of total greenhouse gasses associated with our EVs, appliances, and other products are emitted upfront in their production, before we even start to use them. In this important new book, architect Lloyd Alter, one of our most astute environmental analysts and writers, explains why we also must change our society from one driven by rampant overconsumption into one that adopts common-sense simplicity in our business practices and lifestyles.

— F. Kaid Benfield, co-founder, LEED for Neighborhood Development,
and author, *People Habitat: 25 Ways to Think About Greener, Healthier Cities*

It's critical that we eliminate the annual carbon dioxide emissions from our homes, transit systems, and food systems, but that's not enough—we're also overdue for a careful reckoning of the upfront carbon emissions released in the manufacture of everything we buy. Lloyd Alter explains why we need to ask: Do I really need this product? Will something much smaller serve my needs? and finally, How can this be built without adding carbon to the atmosphere? *The Story of Upfront Carbon* is enlightening and it is essential.

— Bart Hawkins Kreps, co-editor,
Energy Transition and Economic Sufficiency

THE STORY
of
UPFRONT
CARBON

How a Life Of
Just Enough
Offers a Way Out of
the Climate Crisis

LLOYD ALTER

new society
PUBLISHERS

Cover design by Diane McIntosh.
Cover images © iStock (bird—Alex Cooper #1412870805, Retro pile of stuff #1262998444—BrAt_PiKaChU, washer—Animaflora #1291423264, keys—Chimpinski #466931959, cell phone—milosluz #92204900, keyboard—Fian Cahyo Dwi Prasetyo #1480149054, tire—Tanaphong #491683660. Interior image: page 1 © #519748076 / Adobe Stock.

Printed in Canada, March 2024.

Inquiries regarding requests to reprint all or part of *The Story of Upfront Carbon* should be addressed to New Society Publishers at the address below. To order directly from the publishers, please call 250-247-9737 or order online at www.newsociety.com

Any other inquiries can be directed by mail to:
New Society Publishers
P.O. Box 189, Gabriola Island, BC V0R 1X0, Canada
(250) 247-9737

LIBRARY AND ARCHIVES CANADA CATALOGUING IN PUBLICATION

Title: The story of upfront carbon : how a life of just enough offers a way out of the climate crisis / Lloyd Alter.

Names: Alter, Lloyd, author.

Description: Includes bibliographical references and index.

Identifiers: Canadiana (print) 20240329260 | Canadiana (ebook) 20240329341 | ISBN 9780865719927 (softcover) | ISBN 9781771423809 (EPUB) | ISBN 9781550927849 (PDF)

Subjects: LCSH: Carbon—Environmental aspects. | LCSH: Consumer goods—Environmental aspects. | LCSH: Manufacturing industries—Environmental aspects. | LCSH: Climate change mitigation.

Classification: LCC QD181.C1 A48 2024 | DDC 363.738—dc23

Funded by the Government of Canada · Financé par le gouvernement du Canada | Canadä

New Society Publishers' mission is to publish books that contribute in fundamental ways to building an ecologically sustainable and just society, and to do so with the least possible impact on the environment, in a manner that models this vision.

Contents

Acknowledgments

This all started in a Twitter chat with Jorge Chapa and Elrond Burrell, so I have to thank them first. Chris Magwood first introduced me to embodied carbon, "the elephant in the room." Then my fiercest critic, Wolfgang Feist, who continues to challenge me. Lewis Akenji started me on the road to the 1.5 degree lifestyle, and Will Arnold gave me my current three word mantra, "Use less stuff." Rob West and everyone at New Society Publishers gave me this opportunity, but I couldn't do anything without the support of my wife, Kelly. Thank you all.

CHAPTER I

THE LENS OF UPFRONT CARBON

Pick up your phone and feel its weight. It's not much; probably not much different than my iPhone 11 Pro, which weighs 188 grams, or 6.63 ounces. Apple has designed it to be incredibly efficient and run all day on a small battery, so it takes almost no energy to run.

The iPhone is a complicated mix of aluminum, carbon, silicon, cobalt, hydrogen, lithium, tantalum, vanadium, and gold. Materials come from the Democratic Republic of the Congo, Indonesia, Brazil, and China. Metallurgist David Michaud told Brian Merchant, author of *The One Device*, that about 75 pounds of ore were mined to make the phone. Most people could lift that if they had to.

But Apple, one of the few companies to provide the public with a full life cycle analysis showing the carbon emissions of their products, tells us that my phone emits 80 kilograms (176.3 pounds) of carbon dioxide over its lifetime: 13 percent comes from the electricity for operating the phone, 3 percent for transportation, and an astonishing 83 percent from making the phone, the materials that go into it, and the manufacturing. Between manufacturing and shipping, 86 percent of the life cycle carbon is emitted before you open the box. That's 68.8 kilograms, or about 150 pounds—twice as heavy as the ore mined to make the phone.

Lifting your phone is easy, but imagine if it actually weighed 150 pounds. This is serious weightlifting.

The amount of energy that goes into making a product used to be called "embodied energy," defined as "the sum of all the energy required to produce any goods or services, considered as if that energy was incorporated or 'embodied' in the product itself."[1] However, we are in a climate crisis caused by carbon dioxide, methane, and other greenhouse gas emissions, so instead of measuring embodied energy, we started measuring what became known as "embodied carbon." But the dictionary definition of embodied is "include or contain (something) as a constituent part." The carbon is most definitely not a constituent part; it is in the atmosphere.

In a world where we must reduce and eventually eliminate carbon dioxide emissions, this is important. Before you pick up your phone at the store, those carbon emissions have contributed to climate change. They could be considered "now" emissions, compared to "later" emissions, but they are certainly not embodied emissions.

I thought embodied carbon was a terrible name. In a Twitter discussion with New Zealand architect Elrond Burrell, we tried to come up with a better one. Elrond suggested "burped" or "vomited" carbon to make it obvious that they were a giant cloud of carbon emitted during manufacture. Others suggested "front-loaded emissions." Jorge Chapa of the Green Building Council of Australia tweeted, "I also wonder how much people dismissing embodied carbon is the way we talk about it. Instead of embodied carbon, perhaps we should consider renaming it as upfront emissions." I tweeted back, "I think you nailed it!" and added a word, coming up with "Upfront Carbon Emissions."

Writing a year later, author and sustainability provocateur Martin Brown credited me with the coinage:

> Lloyd Alter writing in Treehugger established upfront carbon as a key concept term in addressing the climate emergency. 'Embodied carbon' is not a difficult concept at all; it is just a misleading term.... I have concluded that it should be called upfront carbon emissions, or UCE."

(By the way, Lloyd's article "Let's Rename 'Embodied Carbon' to 'Upfront Carbon Emissions'" is a must-read that also illustrates how Twitter conversations with Elrond Burrell can lead to improved industry thinking.)[2]

I may have given it wings, but in all fairness, it was a discussion among Elrond, Jorge, and myself, and "upfront carbon emissions" is now an accepted term. Jorge Chapa and the World Green Building Council were the first to officially use it in a publication titled "Bringing Embodied Carbon Upfront." Chapa explains why he thinks it is useful:

> We were trying to get funding to do some work on embodied carbon, and while explaining it to a number of funders, about 10 minutes into the conversation one of them stopped us, apologized, and asked a question, "Why do you keep saying embodied carbon is a problem? Isn't embodied carbon good? It's in the product, that's what embodied means, isn't it?" Biggest penny drop I ever had.

However, upfront carbon is not strictly the same thing as embodied carbon, as I will explain later. And whether it's burped, vomited, or just upfront, it is what is going into the air now; it's what is important now; it's the 150 pounds of iPhone upfront carbon that matters in the fight against climate change. When you look at the world through the lens of upfront carbon, everything changes.

When Apple did its life cycle analysis, it attributed 13 percent of emissions to the electricity used to charge the phone based on the average American electricity supply, much of which is still made with coal and natural gas and produces significant carbon emissions. However, if you live in Montreal or Vancouver, where your electricity is generated with water power, that 13 percent drops to almost zero, and the upfront carbon increases as a percentage. The same thing is true if you are driving a Ford F-150 lightning electric pickup truck in Montreal or Oslo where the electricity is low carbon, or you build an all-electric home in Reykjavik: there are no carbon emissions from running the phone, the car, or the house—it's all upfront. As we ramp up renewables and switch to electric vehicles for driving and heat pumps for heating, this leads to what I have called the ironclad rule of upfront carbon:

> As our buildings and everything we make become more efficient and we decarbonize the electricity supply, emissions from embodied and upfront carbon will increasingly dominate and approach 100 percent of emissions.

Everything becomes like your phone with tonnes and tonnes of carbon emissions before you drive the electric car off the lot or step into your new home or unbox a pair of shoes. For products such as your shoes or your sofa, there are no operating emissions; they are almost 100 percent upfront carbon, with just a bit ascribed to maintenance and end of life.

This is why what we make and how much we consume becomes as or more important than how much energy it takes to operate. This is why **sufficiency**, or making and buying just what we need, has become as important as **efficiency**. This is why when you look at the world through the lens of upfront carbon, everything changes.

"Embodied carbon" is doubly confusing because not only is it not embodied, it is not even carbon. Our problem is carbon dioxide, which forms when we burn carbon to generate heat, which happens when a carbon molecule has an exothermic reaction with two molecules of oxygen to make carbon dioxide. So, burning a one-kilogram lump of coal actually has about 3.67 kilograms (8 pounds) of upfront carbon emissions because of the weight of the oxygen.

We also talk about carbon dioxide equivalents (CO_2e), measuring the impact of methane or refrigerants in terms of their effectiveness as greenhouse gases compared to CO_2. It's messy because they are not really equivalent; methane, for example, decomposes in about twenty years, whereas CO_2 stays up in the atmosphere. But for convenience and brevity, when we say carbon, we mean CO_2 or CO_2e, even though what we call carbon is 3.67 times the weight of (solid) carbon.

What Are Upfront Carbon Emissions and Why Are They Important?

When you buy a car, it's easy to find out the fuel economy, the miles per gallon, or, as they do it backward in Canada, the liters per hundred kilometers. It's the law; the Environmental Protection Agency (EPA) mandates the tests on every car. The tests are done to ensure that companies are hitting their targets set by regulation for Corporate Average Fuel Economy (CAFE). The EPA publishes the city and highway fuel economy numbers to encourage the public to compare and buy more efficient vehicles. It's an artifact from when governments were concerned about how much fuel was imported from foreign sources before anyone cared about carbon dioxide emissions.

Today, because we know that carbon dioxide is a greenhouse gas that contributes to global heating, we care a lot about how much CO_2 emissions come out of our tailpipes, which are the car's operating emissions. They are proportionate to the fuel economy; burning a liter of gas emits 2.3 kilograms (5 pounds) of CO_2.[3] That's why governments are promoting electric cars— they have no direct tailpipe operating emissions. They are not emission-free because of the emissions from generating electricity, which is why the EPA conveniently provides a calculator that tells you about the emissions from your electric car depending on the model and where you live based on the cleanliness of your electrical supply and will give you the miles-per-gallon equivalent.[4]

What few companies tell you is what the upfront carbon emissions are—how much carbon dioxide and other greenhouse gases were emitted while actually making the car. There are emissions from making steel, glass, aluminum, and plastics, and in the new electric cars, the stuff that goes into the batteries. There are more emissions from moving these parts around the globe.

When you look at any pie chart showing where carbon emissions come from, these are all attributed to the "industrial" sector, and not to the car. These emissions are considerable, and can be close to the emissions that come out of the tailpipe over the entire lifetime of a gasoline-powered car.[5] An electric car running on clean power is subject to that ironclad rule, and approaches 100 percent upfront carbon.

Upfront carbon emissions are the front end of a life cycle assessment (LCA), a concept that was developed in the early years of the environmental movement and the energy crisis. According to "Life Cycle Assessment: Past, Present, and Future" by Jeroen Guinée:

> The study of environmental impacts of consumer products has a history that dates back to the 1960s and 1970s.... It has been recognized that, for many of these products, a large share of the environmental impacts is not in the use of the product, but in its production, transportation, and disposal. Gradually, the importance of addressing the life cycle of a product, or of several alternative products, thus became an issue in the 1980s and 1990s.

Surprisingly, one of the first to use LCAs was the Coca-Cola Company in 1969, probably to justify the elimination of returnable bottles. According to "A Brief History of Life Cycle Assessment," the study "laid the foundation for the current methods of life cycle inventory analysis in the United States. In a comparison of different beverage containers to determine which container had the lowest releases to the environment and least affected the supply of natural resources, this study quantified the raw materials and fuels used and the environmental loadings from the manufacturing processes for each container."[6]

In his 2008 book *Sustainable Energy Without the Hot Air* David MacKay includes a chapter titled "Stuff," where he discusses the energy required for raw materials (R), production (P), use (U), and disposal (D). Writing about the energy costs of phases R and P, he notes that "These energy costs are sometimes called the 'embodied' or 'embedded' energy of the stuff—slightly confusing names, since usually that energy is neither literally embodied nor embedded in the stuff."[7]

It's one of the earliest references to the embodied energy of stuff, and it's amusing that MacKay noted a decade before I did that the names are confusing. When we talk of embodied carbon, it is even more confusing since it is obviously in the atmosphere and not in the stuff. Many still confuse the terms embodied energy and embodied carbon, even though they are very different.

MacKay includes everything in stuff, such as cars and houses, but it is the building sector that was the first to take the issue seriously.

Today, an LCA is something that most companies can do relatively easily; there are many databases and software programs where you enter the amount of material and multiply that by the CO_2 emissions per kilogram. The programs know the emissions from the power supply where the material is made and the shipping to get it to where it is used. But very few companies reveal the information, even if they have it, possibly because people might be shocked.

As Paolo Natali of RMI wrote about electric cars:

> The truth is that the accumulated carbon footprint of materials in a newly bought gasoline-fueled car is the same order of magnitude as the footprint of its lifetime fuel consumption—so by buying an electric vehicle and securing green electricity, you are only part of the way through abating your car's total carbon footprint.
>
> What can we do to change this? Because what is out of sight is often out of mind, the first step is to calculate and communicate the CO_2 emissions that are embedded in produced goods. Until people know the CO_2 footprint of the products they're using, it will be impossible for them to demand lower-carbon goods.[8]

But the data aren't there for us to calculate the upfront carbon emissions of a car. One of the first to try was Mike Berners-Lee, author of *How Bad Are the Bananas*, one of the inspirations for this book. Back in 2010, he described in the *Guardian* how hard it is; first you have to draw a fence around it:

> To give just one simple example among millions, the assembly plant uses phones and they in turn had to be manufactured, along with the phone lines that transmit the calls. The ripples go on and on for ever. Attempts to capture all these stages by adding them up individually are doomed from the outset to result in an underestimate, because the task is just too big.

They then did what they call an input-output analysis, determining the total consumption of different materials that the auto industry consumed, the emissions from making those materials, and then divided it by the total amount of money spent on cars, and came up with the number 720 kilograms (1,587.3 pounds) of CO_2e for every thousand UK pounds spent on a car.

My first thought was that this is silly; you can have a Toyota and a Lexus that are identical under the skin but have very different prices, but Berners-Lee calls it "a reasonable ballpark estimate." There are likely much better numbers and approaches that one would take today, but without hard data from the manufacturer, it is impossible to know the true number.

For example, the most popular vehicle in North America is the Ford F-150 pickup truck. A few years ago, Ford started making the truck out of aluminum instead of steel to make the vehicle lighter and get better fuel economy—or, if you are a cynic like me, make it even bigger. Virgin aluminum has a vastly higher carbon footprint than steel, but Ford doesn't tell us whether they are using virgin or recycled aluminum with 95 percent fewer emissions. They make a very big deal about recycling their pre-consumer scrap aluminum, which is greenwashing; every company does that. Nobody will throw away

30 to 40 percent of the aluminum left after stamping out a part. But they don't tell us where the original aluminum sheet comes from, and without knowing that, you can't even ballpark a number.

Without transparency and openness, we will never have accurate information, and with the automotive industry, we will never have transparency. They want to sell big high-end vehicles, which have massive upfront carbon emissions no matter what they run on, and are directly proportional to the size and weight.

Why We Are Fixated on Energy, Not Carbon

In October 1973, the Organization of Arab Petroleum Exporting Countries declared an oil embargo aimed at nations that supported Israel in the Yom Kippur War. The price of oil tripled, and governments worried about their dependence on foreign energy sources. To reduce energy consumption, speed limits were lowered, efficiency standards for cars were introduced, and building codes were tightened.

In 1977, President Jimmy Carter called dealing with the energy crisis "the moral equivalent of war," declaring that "the cornerstone of our policy is to reduce the demand through conservation. Our emphasis on conservation is a clear difference between this plan and others which merely encouraged crash production efforts. Conservation is the quickest, cheapest, most practical source of energy." He called for the insulation of 90 percent of American homes and all new buildings, solar energy, and smaller cars. "Those who insist on driving large, unnecessarily powerful cars must expect to pay more for that luxury." In what today sounds like a discordant note, Carter also called to "increase our coal production by about two-thirds to more than 1 billion tons a year."[9]

In the forty-five years since that speech, governments have come and gone, but the preoccupation with energy has not. Jimmy Carter could call for more coal production because he was dealing with energy consumption, not carbon emissions, which are two very different problems. For forty-five years, we have measured miles per gallon or energy consumption of our homes and businesses because using energy meant burning fossil fuels. We used vast amounts of energy to boil rocks in Alberta to achieve energy independence. We would spray our homes with polyurethane foam insulation or wrap them in Styrofoam to reduce our gas consumption because we worried about how much fuel it took to run things, what became known as operating energy.

The energy that it took to make things wasn't seen as a big issue back in the energy crisis days; major industrial processes such as making electricity, steel, or concrete used coal, not oil, and it was not imported, so it didn't matter. This was not an energy crisis, but an oil crisis. And while Jimmy Carter tried to address the demand side to reduce consumption, his successor, Ronald Reagan, had other ideas, and worked the supply side. As Indrajit Samarajiva writes:

> The response to pressure from oil-producing countries could have been to use less oil, but no. The response was to produce more oil. The policy running all the way through Obama and Trump and Biden has been oil independence, not independence from oil.[10]

Ronald Reagan gets all the credit for ending the oil crisis by deregulating the price of crude oil in 1981 and letting the industry drill offshore—both the North Sea and Prudhoe Bay in Alaska came online—and just about anywhere else, but Jimmy Carter set him up for success. The average fuel economy of cars went from 20 miles per gallon (mpg) in 1978 to 28 mpg in 1985, while homeowners in droves switched from oil to gas and electricity for heating.[11] There were also geopolitics in play—Iraq was at war with Iran, and both were dumping oil on the market to pay for it. There are some who say that Reagan made a deal with Saudi Arabia to pump more oil and drop the price to destroy the Soviet Union, which then, as now, depended on fossil fuels for most of its income.[12] Oil went from being scarce to being a glut, and the price collapsed.

Oil got cheap and stayed cheap for a long time, making it very hard to get anyone to care much about using less of it. And as codes and regulations kept making cars and buildings more energy efficient, they just kept getting bigger because if you are a consumer, why not? It doesn't cost you any more to operate them. But as they got both bigger and more efficient, people started noticing that more and more energy was being used to make things in relation to how much it took to operate them. Academics called this "embedded energy" or "embodied energy." Here's an interesting definition from the *Encyclopedia of Energy*, published in 2004:

> Embodied energy, or "embedded energy," is a concept that includes the energy required to extract raw materials from nature, plus the energy utilized in the manufacturing activities. Inevitably, all products and goods have inherent embodied energy. The closer a material is to its natural state at the time of use, the lower its embodied energy. Sand and gravel, for example, have lower embodied energy as compared to copper wire. It is necessary to include both renewable and nonrenewable sources of energy in an embodied energy analysis.[13]

Note that they do not include the emissions from the schlepping of the heavy sand and gravel, nor the difference between renewable and nonrenewable sources of energy. That's the preoccupation with energy, not carbon, writ large.

Carbon Takes Command

The energy crisis faded in the onslaught of "unconventional" resources, as hydraulic fracturing (fracking) unleashed vast quantities of oil and more natural gas than we could use. In Canada, the oil companies got better at squeezing oil out of the rocks of what became recognized as the fourth-largest petroleum reserve in the world.[14] There was a brief flurry of worry about "peak oil," but companies kept finding more of the stuff, especially gas, and we kept burning more of it as economies grew along with the cars and houses. The party continues because it's what drives our economies; as Vaclav Smil noted in his book *Energy and Civilization: A History*:

> By turning to these rich stores we have created societies that transform unprecedented amounts of energy. This transformation brought enormous advances in agricultural productivity and crop yields; it has resulted first in rapid industrialization and urbanization, in the expansion and acceleration of transportation, and in an even more impressive growth of our information and communication capabilities; and all of these developments have combined to produce long periods of high rates of economic growth that have created a great deal of real affluence, raised the average quality of life for most of the world's population, and eventually produced new, high-energy service economies.

In all these years after 1973 when we worried about our "energy crisis"—which was really a politically motivated gasoline crisis—scientists were beginning to understand that we were going to find ourselves in a carbon dioxide crisis. Back in 1981, the warmest year on record at that time, even the oil companies could see this coming.

M.B. Glaser, manager of the Environmental Affairs Program at Exxon, told his bosses in 1981 that "our best estimate is that doubling of the current concentration could increase average global temperature by about 1.3 degrees Celsius to 3.1 degrees Celsius." Glaser also noted—in 1981!—that "mitigation of the 'greenhouse effect' would require major reductions in fossil fuel combustion."[15] Needless to say, this report never saw the light of day.

There was also solid evidence that chlorofluorocarbon (CFC) from leaky fridges and air conditioners were causing atmospheric changes, including the enlargement of the "ozone hole." Somehow, the nations of the world got together to ban Freon and regulate CFC with the Montreal Protocol of 1987,

which became the definitive model for international cooperation; it demonstrated that even in the face of industry and ideo-
logical objections to regulation, agreements could be reached
that could come up with market-oriented mechanisms for solv-
ing the problem.[16]

In 1988, in the face of an increasing pile of evidence about
the dangers of greenhouse gases, the Intergovernmental Panel
on Climate Change (IPCC) was founded by the United Nations
Environment Program (UNEP) and the World Meteorologi-
cal Organization (WMO). The intent was to examine the data,
understand the dangers, and determine solutions based on
the Montreal model of international cooperation, agreement,
and action.

Shortly thereafter, the fossil fuel and automotive industries founded the Global Climate Coalition, whose mission,
according to Spencer Weart in his book *The Discovery of Global
Warming*, was to disparage every call for action against global
warming.

> This effort followed the pattern of scientific criticism,
> advertising, and lobbying that industrial groups had
> earlier used to cast doubt on warnings against ozone
> depletion, acid rain, and other dangers as far back as
> automobile smog and leaded gasoline. But the most
> obvious model was the long-sustained and dishonest
> campaign by the tobacco industry, which had shortened
> many millions of lives by persuading people that the
> science of smoking was controversial.[17]

This worked remarkably well, delaying the implementation of
agreements literally by decades. There are still what we used to
call climate skeptics, then deniers, and now climate arsonists
in conservative parties around the world. Even governments
that are trying to make the kind of changes needed to honour
the agreements they already made make little progress because
of entrenched interests. It's also why it has been so difficult

even to get people to understand the issues of carbon, let alone regulate them; there is so much disinformation and wishful thinking, and people still are thinking about energy, not carbon. The fossil fuel industries like it that way; they have lots of energy to sell.

If people were not talking enough about carbon before the war in Ukraine, there has not been a peep about it since. Energy has been back on the front burner since Russia cut off supplies of natural gas to Europe in 2022, causing massive disruption in economies around the world and a big spike in fossil fuel prices. People and politicians are not worrying about carbon emissions these days; Alberta sees a natural gas gold mine, and Britain's energy minister says, "We need to be thinking about extracting every last cubic inch of gas from the North Sea." Jacob Rees-Mogg may yet send children back into the coal mines.

The IPCC Does Not Say We're Doomed

The IPCC has been churning out Assessment Reports since 1990, and has now completed six cycles. In 1992, the United Nations Framework on Climate Change (UNFCC), a "universal convention that recognizes the existence of climate change due to human activity," was signed in Rio de Janeiro by nations that became known as the Conference of the Parties (COP), which has been partying annually ever since to determine collective responses to the IPCC reports.

In 2015, the COP signed the Paris Agreement, where Party nations would come up with plans for climate action called Nationally Determined Contributions (NDC) required to keep global heating under 2°C.

In 2018, a special report was published titled "Global Warming of 1.5°C," which included what was described by author Daniel Yergin as "one of the most important sentences of last few centuries. It has provided an incredibly powerful traffic signal to tell you where things are going."[18] It is certainly not written in words to stir one's soul:

> In model pathways with no or limited overshoot of 1.5°C,
> global net anthropogenic CO₂ *emissions decline by about*
> *45% from 2010 levels by 2030 (40–60% interquartile range),*
> *reaching net zero around 2050 (2045–2055 interquartile*
> *range).*

This sentence has generated a thousand graphs of emission gaps, showing how we are failing to cause those emissions to decline. It has also been the basis of a thousand pledges to reach net zero by 2050, while studiously avoiding doing very much at all right now or even by 2030.

But Daniel Yergin is wrong. The sentence is not a driver of carbon reductions or even a target; it is an excuse for maintaining the status quo and pushing the problem down the road.

I believe that the sentence for the ages is in the Working Group I's Sixth Assessment Report:

> To limit global warming to 1.5°C above pre-industrial
> levels with either a one-in-two (50%) or two-in-three
> (67%) chance, the remaining carbon budgets amount
> to 500 and 400 billion tonnes of CO₂, respectively, from
> 1 January 2020 onward.

It is followed with the note: "Currently, human activities are emitting around 40 billion tonnes into the atmosphere in a single year."

The key difference between Daniel Yergin's favorite sentence and mine is that Yergin thinks that the carbon budget is a budget and not a ceiling. Think of it as a person's fixed retirement nest egg. If every penny matters, then you would cut your spending fast and hard. But instead, people are saying, "I'll buy that yacht I have always wanted right now, but I'll eat cat food in 2030," while thinking, "I'll be long dead in 2050." Mine has a ceiling that is based on hard science: global heating is proportional to the amount of carbon dioxide and equivalents in the atmosphere, and every molecule of CO₂ that we add to the atmosphere is subtracted from that budget. As Kimberley Nicholas noted in her book *Under the Sky We Make*:

> Because carbon is essentially forever, the carbon budget is forever too. If I use up more than my share, this leaves less space for you. This is true today across places: between rich and poor countries and between high- and low-emitting individuals. This tug-of-war is also true stretching across time: between previous generations, those of us alive today, and all humanity to follow.

It has also become fashionable to, as *The Economist* magazine put it, "Say Goodbye to 1.5°C." Or as the environmental website Grist asked, "The world's most ambitious climate goal is essentially out of reach. Why won't anyone admit it?"[19] Shannon Osaka of Grist wrote that after cherry-picking a line from the Working Group III: Mitigation of Climate Change Assessment Report that came out in April 2022: "Hidden on page 25 of the 'Summary for Policymakers' was an even grimmer note: That even in the IPCC's most optimistic models, the chances of holding global warming to less than 1.5 degrees Celsius (2.7 degrees Fahrenheit)—compared to the pre-industrial average—is only around 38 percent.... For all intents and purposes, the 1.5-degree threshold has already passed. We just don't know it yet."

All I can say is that I must be reading a different assessment report because I couldn't find that on page 25, and found that the rest of the document laid out a path to staying under 1.5 degrees. It's all in the title: "Mitigation of Climate Change" through greater efficiency, increased use of renewables, and "demand mitigation."

> Demand-side mitigation encompasses changes in infrastructure use, end-use technology adoption, and socio-cultural and behavioural change. Demand-side measures and new ways of end-use service provision can reduce global GHG emissions in end-use sectors by

40–70% by 2050 compared to baseline scenarios, while some regions and socioeconomic groups require additional energy and resources. Demand-side mitigation response options are consistent with improving basic well-being for all.

Demand-side mitigation is about asking, "What is enough?" Do we need a pickup when an e-bike might do? Would that leave enough metal and lithium that we could give e-bikes to people who would have their lives improved by them? How big an apartment do I need? How many pairs of shoes? Do I need the latest iPhone? Another word for it is **sufficiency**.

Working Group I studied the physical science basis and concluded, "It's real!" Working Group II looked at the impacts and vulnerability and concluded, "It's bad!" The key takeaway:

Global warming, reaching 1.5°C in the near-term, would cause unavoidable increases in multiple climate hazards and present multiple risks to ecosystems and humans (very high confidence). The level of risk will depend on concurrent near-term trends in vulnerability, exposure, level of socioeconomic development and adaptation.

Working Group III concludes: "We can fix it!" Climate journalist Amy Westervelt nailed it in her analysis in the *Guardian*: "The report made one thing abundantly clear: the technologies and policies necessary to adequately address climate change exist, and the only real obstacles are politics and fossil fuel interests."[20]

OK Doomer

About a decade ago, climate journalist Dana Nuccitelli described the five stages of climate denial as the IPCC released its fifth round of reports. They were:

Stage 1: Deny the problem exists. We are well past that now, although a few flat-out deniers still exist in comments sections of newspapers.

Stage 2: Deny humans are the cause. There are still a few of these about, still blaming sunspots and claiming that the earth goes through natural cycles.

Stage 3: Deny it's a problem. More CO_2 means more plants! More warming means more Canada!

Stage 4: Deny we can solve it. It's too expensive; it will hurt the poor; it will trash the economy. This is the popular one right now with Lomborg, Shellenberger, and other eco-modernists.

Stage 5: It's too late. When Nuccitelli wrote these, he noted that "few climate contrarians had reached this level." Today, the world is full of what author and climate scientist Michael Mann called "doomists."[21]

> Exaggeration of the climate threat by purveyors of doom—we'll call them "doomists"—is unhelpful at best. Indeed, doomism today arguably poses a greater threat to climate action than outright denial. For if catastrophic warming of the planet were truly inevitable and there were no agency on our part in averting it, why should we do anything? Doomism potentially leads us down the same path of inaction as outright denial of the threat. Exaggerated claims and hyperbole, moreover, play into efforts by deniers and delayers to discredit the science, posing further obstacles to action.[22]

Hannah Ritchie of Our World in Data recently raised the same point, suggesting that doomers were worse than deniers.

> Climate deniers want us to *choose* to do nothing; that it's not a problem and doesn't require any action. Climate doomers tell us that we *don't even have a choice* to do something; we're already screwed and it's too late to act. Follow either and we end up in the same place of in-action. That's a place that we can't afford to be.[23]

Author Jonathan Franzen is a key "doomer," as I prefer to call them, telling Australian radio that "We literally are living in end times for civilization as we know it.... We are long past the point of averting climate catastrophe." Brynn O'Brien, executive director of the Australasian Centre for Corporate Responsibility, responded and is quoted in the *New Statesman*:

> The only people who fall for it are rich white people who think they will be spared until everyone and everything else is gone. His position is unscientific, morally careless (at best) and politically blinkered. Things are very bad and will get much worse. But scientifically and politically there are still so many choices we can and must make to avoid all-out catastrophe, to avoid "end times."[24]

The doomers were out in force when the annual Emissions Gap report from the United Nations Environment Programme was released just before COP27. Everyone piled on one sentence: "As climate impacts intensify, the Emissions Gap Report 2022 finds that the world is still falling short of the Paris climate goals, with no credible pathway to 1.5°C in place." They all then proceeded to totally ignore the two sentences which directly followed: "Only an urgent system-wide transformation can avoid an accelerating climate disaster. The report looks at how to deliver this transformation through action in the electricity supply, industry, transport and buildings sectors, and the food and financial systems." The report then laid out what UNEP Executive Director Inger Andersen called a "root and branch transformation" of our economies and societies, with many of the same "demand-side mitigations" called for in the IPCC report. The emissions gap between where we are and where we have to go can be closed with reductions in demand, living in smaller spaces, switching to lower-emitting modes of transport such as bikes and public transit, eating less meat, and building better buildings. These are again all about sufficiency.

But the doomers have a point when they quote the report. In his book *I Want a Better Catastrophe*, Andrew Boyd tries to put the best spin on being a doomer, suggesting that people and organizations may be putting their best spin on bad news.

Inger Andersen calls for a "root and branch transformation," when she knows it is not happening. Is she just telling people the most hopeful version of the truth, as Boyd suggests? Boyd describes the process:

> Try to be as positive and pragmatic as you can be. Tell the best possible version of events. Focus on the promise and potential of the moment. And fight like hell and hope for the best.

This is the position I have taken: that we know what to do. Otherwise I wouldn't have much of a book here. It's hard sometimes when we see so little progress. As I write this, I am trapped inside a cabin in the woods because the air outside is toxic, full of smoke from forest fires in Quebec. But I am still positive and pragmatic. I am not alone; Professor Kevin Anderson of the Tyndall Centre for Climate Change Research, perhaps one of the direst climate experts, said in a recent interview:

> We are going to fail. We are going to go to 3 or 4 degrees Centigrade of warming and we will have to live through or die from all of the repercussions that that will have. That is a terrible prospect and one that I think we have to try everything to avoid. But the message of hope, if there's any thread of hope in this, is that it is a choice to fail...we can choose a different way out of this. Now whether we can still hold to 1.5, it looks incredibly unlikely to me. But incredibly unlikely doesn't mean it's impossible. It is only impossible if we don't try.[25]

And that is why we are here, to fight like hell and hope for the best. And as the UNEP Emissions Gap report notes, we have to do it now, with every decision we make, to avoid what's called "lock-in":

> Decisions made today can define emissions trajectories for decades to come. For example, a building lasts 80 years on average; a coal-fired power plant 45 years;

a cement plant 40 years. Pipelines and gas connections create decade-long dependencies. Interventions can also lock in behavior and policies that reinforce incumbent systems. Actions today that lock in a high-energy and high-carbon future for decades must be avoided, including avoiding new fossil fuel infrastructure for electricity and industry, car-centered city or regional planning, and inefficient new buildings. These actions do not always result in immediate emission reductions, but are fundamental for the long-term transition.

Lock-in is disastrous when we are talking such small numbers remaining in the carbon budget and so little time to reduce emissions. But lock-in doesn't just happen at the industrial level with highways and cement plants—it happens at the personal level with the decisions that we make in our own lives. In 2014, when renovating my own home, I bought a new gas boiler to pump hot water to our hundred-year-old radiators. Heat pumps were just too expensive at the time, so I locked myself into gas. People do this every day when they buy new cars or new houses in the suburbs; they are locked into fossil fuels for years to come.

Every giant new pickup truck I see in my neighborhood is lock-in writ large. It seems there should be a Stage 6 of climate denial: "It's happening, it's real, it's someone else's problem, and I don't care."

This is why socio-cultural and behavioral change as well as demand-side mitigation are so critically important. Just as every kilogram of carbon we add to the atmosphere is subtracted from the carbon budget, every new lock-in ensures that we keep subtracting it for years to come.

The Building Industry and Upfront Carbon
The embodied emissions from making the stuff that goes into new construction, concrete, steel, glass, aluminum, and putting it together are responsible for 11 percent of global carbon

emissions. With the IPCC calling for reductions in carbon emissions of 65 percent by 2030 to stay under 1.5°C, it became clear that the 11 percent had to be reduced. Architecture 2030 issued a challenge where "The embodied carbon emissions from all buildings, infrastructure, and associated materials shall immediately meet a maximum global warming potential (GWP) of 40% below the industry average today," increasing to 65 percent by 2030 and to 100 percent emissions by 2050.[26] Other organizations hopped on board and governments, particularly in Europe, started demanding data.

Only a small percentage of the profession is doing anything more than paying lip service to reduce upfront carbon, but it is a big enough industry that tools have been developed to measure it. Definitions have also become more precise.

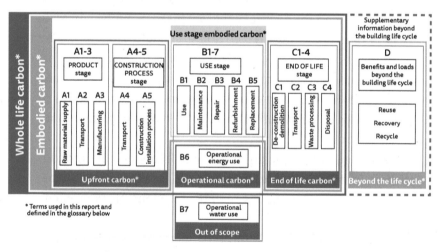

Carbon stages via World Green Building Council

Embodied carbon is now considered to be "the greenhouse gas emissions arising from the manufacturing, transportation, installation, maintenance, and disposal of building materials"[27]—basically, everything in the full life cycle of the building, not including the operational carbon (the carbon emitted while operating the building). As shown on this chart from the World Green Building Council, "upfront carbon" is now considered to be the emissions from the product stage, the raw materials, transport to the factory, the manufacture and construction stage, the transport to the site, and the installation. As an aside, this chart and the report it came from was the first to use the term "upfront carbon," and the term is coming into common use in Europe and Canada.

There are now many tools for measuring embodied and upfront carbon in buildings, and life cycle analyses are now often required by some municipalities that have set targets to reduce upfront carbon emissions. These are usually cradle-to-grave studies, but this book focuses on the upfront section, from cradle to installation. And even in the building world, where the issue of embodied carbon is probably as sophisticated as it gets, nobody is quite sure how accurate they are.

Some people are not even sure we are measuring the right thing. Where the World Green Building Council and I include the construction process stage in upfront carbon, carbon pioneer Chris Magwood uses another metric, "Material Carbon Emissions," which only includes the product stage, A1–3 stages, in the table. I asked him why he didn't include A4 and A5, and he told me, "Two reasons they weren't included: They are much less significant than might be expected (3–6 percent of total emissions), and it's impossible to estimate them accurately." He has a point; the transport emissions could be all over the place, depending on the distance from the factory or warehouse. On the other hand, including it has encouraged builders to take delivery of their mass timber by rail rather than truck. And for our purposes, in discussing the carbon footprint of everything, it becomes more important.

The Carbon Footprint of Everything

Every kind of stuff has emitted varying amounts of upfront carbon.

In 2022, I assigned a strange project to my sustainable design students at Toronto Metropolitan University: They were each to pick an object and figure out the upfront carbon from the sourcing of the materials, the manufacture, and the delivery. My intent was that they would get an understanding of the importance of upfront carbon in everything we use every day. In retrospect, there were some serious flaws in the assignment— by leaving the selection of the objects to the students, there was a very strange mix of everything from a container of marijuana to a shoebox. I didn't make it clear what should be included, so a student calculating the carbon footprint of a pencil found it was dominated by her drive from home to the Staples store and back. The carbon footprint of transportation befuddled many of them; a student studying a watermelon included a transport number so high that I could have had it delivered by helicopter. The numbers could be wildly inconsistent; three students studied a cup of coffee and came up with 16 grams (0.6 ounces) (too low), 120 grams (4.2 ounces) (a bit high), and 25.4 kilograms (55.9 pounds) (she flew the beans from Ethiopia).

Perhaps the biggest failure of the assignment was that I hadn't done the research for this book yet, and did not lay out a straightforward and consistent format often used in life cycle assessments. It comes down to three main components:

1. Materials. You list all the materials going into the product by weight. Databases are available to determine the amount of carbon released from each material—the carbon intensity.

   Materials emissions (Kg CO_2e) = component quantity × carbon intensity

2. Manufacturing. You add to that the energy involved in making the product in Kilowatt-hours—not an easy calculation—and determine how clean the energy supply is where it was made, another number that can be wildly variable.

   Manufacturing = Energy (kWh/process) × regionally specific grid factor

3. Transport. You calculate the distance times the mode of transport—truck, ship, or plane.

   Transport = distance × weight × mode in CO_2 per kg per km

This is not too difficult for a relatively simple product made mostly under the control of the manufacturer, but it is more complicated if it is a complex product that is made from parts that come from all over the world and might have final assembly in China. If the supply chain gets complex, then so does the life cycle assessment.

This methodology was developed for the construction industry, which consumes a lot of material, but it applies to anything from shoes to motorcycles. I should have used it with my class instead of leaving the methodology up to them.

Many companies issue single-company, single-product Environmental Product Declarations (EPD), which are prepared according to ISO or European standards and reviewed by third parties, listing the environmental characteristics and the upfront carbon. Many companies do not bother to do them, because they are expensive or could reveal proprietary information.

A good EPD can be an interesting read, defining what the product is, what it is made of, how it is made, and what its environmental impact is in terms of global warming potential, ozone depletion, acidification, eutrophication, and more. They can also be complicated and incomprehensible. It has also often been said that "not all EPDs are created equal," and much depends on the quality of the life cycle assessment.

Most are primarily related to construction products; it is hard to find them for shoes or motorcycles.

Nonetheless, many of the results my students came up with were fascinating and illuminating. Almost half of the footprint of the Nike shoebox came from the inkjet printing of a red Nike graphic on the box, which is seen for a few minutes between when someone takes delivery of their shoes and disposes of the box. A 5.71-gram (2-ounce) gold necklace that serves no useful purpose is responsible for the emissions of 163 kilograms (359.3 pounds) of CO_2, a number so extraordinary that my student checked it against four other references. That's a ratio of 28.669 kilograms (63.2 pounds) of CO_2 per gram of gold.

According to consumerecology.com, the fabrication of a standard 9 × 9 × 9" ULINE cardboard box creates 270 grams of CO_2. In order to find how many grams of CO_2 were produced per square inch of the box, I created an unfolded template for it and calculated that it would produce .41 g of CO_2 per square inch of cardboard.

I used the size of a standard shoebox, which is 13 × 8 × 6", as my template size for the purpose of this research. Once again, I unfolded this box in order to find out how many square inches it accounts for. I then multiplied this number by the CO_2 per square inch number that I had found from the 9 × 9 × 9" ULINE box. This size box ended up producing 299.3 g of CO_2.

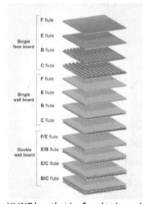

The ULINE box that I refered to is made of more substantial cardboard than the shoebox. I found that the shoebox is made of single wall cardboard, which is thin in comparison to the double wall board of the ULINE box. I decided to split the CO_2 number from my previous calculation in half, since the single wall cardboard occupies half the volume of the double wall.

I also wanted to account that most shoeboxes are printed on. I found that inkjet printing a 8.5" × 11" page produces 6 g of CO_2, according to ezeep.com. Based on this information, I found how much CO_2 inkjet printing produced per square inch and then multiplied this number by the square inches of the shoebox, which gave 46.72 g of CO_2.

Calculations at a Glance

1. 9 × 9 × 9" box (657 square inches) = 270 g CO_2
2. 270 g CO_2/657 square inches = .41 g CO_2 per square inch of cardboard
3. 13 × 8 × 6" box (730 square inches) × .41 g = 299.3 g CO_2
4. 299.3 g CO_2 (double wall cardboard box)/ 2 = 149.65 g CO_2 (single wall box)
5. 8.5 × 11" (93.5 square inches) paper inkjet printed = 6 g CO_2
6. 6 g CO_2/93.5 square inches = 0.064 g CO_2 per square inch printed
7. 13 × 8 × 6" box (730 square inches) × 0.064 g CO_2 = 46.72 g CO_2
8. 46.72 g CO_2 (ink) + 149.65 g CO_2 (cardboard) = 196.37 g CO_2 total

Resources

*https://consumerecology.com/carbon-footprint-of-a-cardboard-box/
*https://www.ezeep.com/co2-neutral-printing/#:~:text=A%20laser%20printer%20
  that%20prints,produces%20approximately%201g%20of%20CO2.

Others demonstrated the complexity and impact of the global supply chain. Transport by ship has a very low carbon footprint per gram of stuff moved; it barely registers on my favorite submission of a cup of tea. But the quantity of stuff moved is so huge that the aggregate effect of shipping emissions is significant.

But whether it was a cigarette (14 grams or .5 ounces), a T-shirt (5 kilograms or 11 pounds), or a pair of jeans (21 kilograms or 46.3 pounds) it became clear to every student that everything we use has its own upfront carbon footprint. And the lesson of the exercise was clear: When I asked on the exam what to do about this, there were some who said, "buy local." But almost every student responded with something like "buy less stuff," "make things last longer," and "reduce waste."

When you get down to the gritty details of making stuff, there are only so many things you can do to squeeze out the carbon emissions. You might take 20 percent out of it if everything was absolutely as carbon efficient as possible, but you still have all these steps, all this material, and in this case, going

What is the Total Embodied Carbon in a Cup of Black Tea?

Cultivation
CO_2e: 0.0135 kg (13.5 grams)
• Fertilizers (rapeseed meal), plastic basket, petrol (pruning), water (assume rainfall)
• Approximately 4 kg of fresh tea leaves go into 1 kg of dried tea

Processing
CO_2e: 0.0015 kg (1.5 grams)
• Withering, rolling, fermentation, drying, baking, sorting
• A single tea bag has around 2 grams of processed, dried tea

Packaging
CO_2e: 0.00842 kg (8.42 grams)
• Standard packaging (domestic use in storefront), abroad packaging (for shipping overseas)
• Choosing loose leaf tea can save up to a third of the CO_2e produced

Transportation
CO_2e: 0.0004 kg (0.4 grams)
• Sea freight (shipped abroad direct from China to Canada), truck transport to storefront, walking to purchase
• Transportation varies based on the path

Consumer Use
CO_2e: 0.0011825 kg (1.1825 grams)
• Electric kettle, water (250 mL)
• Electricity use is lower in places where hydropower is used
• Tap water usage is lower in places with readily available sources and consumes less than bottled water

Disposal
CO_2e: –0.002 kg (–2 grams)
• Packaging (either sent to landfill or 80% recycled, steeped tea leaves removed from the bag and composted
• Tea plants save about 30 grams of CO_2 per cup

**Total CO_2e:
0.0230025 kg
(23.0025 grams)**

Credit: Eira Roberts.

right back to the fields where the cotton is harvested. That's a 4-kilogram saving, if you can find it. But not buying a new pair of jeans because you really didn't need them or you keep wearing the ones you have, that saves 20 kilograms (44 pounds). So, in this example, sufficiency—not buying a new pair of jeans—is five times as effective as the imagined numbers for efficiency. This is true of jeans and almost anything else.

The Unbearable Heaviness of Carbon

One of my students analyzed a sheet of paper for her upfront carbon project. She came up with 558 grams (19.7 ounces) for tree harvesting, pulping, and paper making, and then came up with 27,634 grams (974.8 ounces) of CO_2 emissions for transportation across the country, which is off by a factor of about ten thousand. She was not alone in having trouble visualizing how much CO_2 weighs; it is a gas, and we don't think of gases as having weight.

Many have tried to express CO_2 as volume. The CAKE motorcycle company hung one of their bikes in a giant cube, enclosing the 1,186 kilograms (2,614.7 pounds) of upfront emissions from making one of their bikes. But volume changes with pressure, and I am not certain that people understand that. Bubbles of gas have been used as an image with the explanation that "at standard pressure, and 59°F, a metric ton of carbon dioxide gas would fill a sphere thirty-three feet across." It is too variable.

We started this book by noting that my iPhone had a CO_2 weight of 80 kilograms (176.4 pounds). My original plan was to measure the carbon footprint of everything and put it in units that people could understand, like Wile E. Coyote's ACME anvils of 100 kilograms (220.5 pounds) each. The MacBook Air I am writing on would be presented as eight anvils. In "The Love Song of J. Alfred Prufrock," T. S. Elliot wrote, "I have measured out my life with coffee spoons." I thought this might be a good unit to use; I would call it a Prufrock. But I weighed a coffee spoon, and it was only 26 grams (0.9 ounces). That's a bit light.

Others have tried using carbon dioxide equivalents. Will

Arnold of the British Institution of Structural Engineers has tried to educate the construction industry about embodied carbon by comparing it to the kinds of things we have all been talking about for years: your flight to Europe is a tonne of carbon! A year's worth of meat is 2 tonnes! Driving your car is 3 tonnes per year! He thinks this makes it easier for people to understand the scale of the problem, writing:

> What can you do about it? Well, as an example, if you agree to columns that are closer together, limit the number of storeys, and allow the engineer to spend more time optimising the structure, you might halve it. That's a saving of 3,000 tonnes of embodied carbon…twenty lifetimes of veganism!

I thought this method might work, and considered measuring everything in terms of hamburgers, noting of one construction project that "switching from concrete to cross-laminated timber reduced upfront carbon emissions by the equivalent of 53 people-years or 323,000 hamburgers."

There are so many ways to visualize CO_2; a tonne of it could be a 33-foot bubble, 38,411 Prufrocks, 8,200 kilometres (5,095 miles) of driving, one flight to Europe, 100 hamburgers, 166 trees absorbing CO_2 per year, or ten standard ACME anvils. However you measure it, it is far heavier and bigger than most people imagine.

Transparency from Shoes to Motorcycles

Some companies are not burying their life cycle assessments, but loudly promoting them. Allbirds, which calls itself a sustainable shoe company, says, "To hold ourselves accountable in our mission to lower our carbon footprint, we're measuring our impact and putting our results on display."[28] They use all-natural materials wherever possible, which is not uncommon. However, they also developed their own detailed life cycle assessment tool with third-party environmental consultants. Allbirds notes:

We believe that natural materials, in contrast to petroleum-based materials, have the potential to act as carbon sinks through improved practices such as regenerative agriculture. Although natural materials, like wool, are not always low carbon from the start, they have the potential to be, and Allbirds is committed to supporting research and development to realize this potential.

Allbirds notes that they didn't have much information to work with or other companies to compare data with. They found a 2012 study from MIT, which concluded that the average "standard sneaker" had a carbon footprint of about 14.11 kilograms (31.1 pounds) of CO_2e per pair; the Allbirds average in 2021 was 8.76 kilograms (19.3 pounds).

"Limited research exists on the environmental impact of footwear production—there are few LCAs to begin with, and even fewer are transparent about detailed methodology and assumptions." It's not easy. "When we use a range of data sources, there can be discrepancies in the scope and methodology. We do our best to ensure that values from different sources are comparable, though sometimes we are unable to confirm."

However, they are thorough and transparent and list the carbon footprint of every shoe they sell, along with a complete list of materials. They are soon launching a shoe that is almost fossil-fuel-free, with plastic substitutes made from sugar cane, castor oil, and eucalyptus trees.

CAKE is a Swedish electric motorcycle company which recognized that the upfront carbon emissions matter as much as operating carbon, with words you do not hear from the likes of Ford and GM. "Going fossil free isn't just about how things are powered, it's about removing fossil fuels from how things are sourced, made, transported, and assembled. So while electric vehicles are a great start, we need to go further."

The problem CAKE had was that a motorcycle is different than a pair of shoes; they had to track a hundred assembled components made up of two thousand parts coming from all over the world. "Every single component starting from the bike's frame to a single screw's origin and detailed manufacturing plan was documented. A distinction was drawn between the data provided by the producer and the generic data offered by various databases."

In the end, they essentially gave up. They found it almost impossible to get primary data because of the complexity of the global supply chain and found trying to track it to be "a sobering process." Sixty-two percent of the bike's footprint came from the motor, battery, controller, brakes, and suspension, all from suppliers they had little control over. So, they are going after the big chunks that they can control: the aluminum, steel, plastic, and rubber that they work with in the factory. There is not much that they can do, but they are trying.

But the most important result of the exercise is that people are learning about upfront carbon. Their motorbike emitted 1,186 kilograms (2,614,7 pounds) of upfront carbon; they say a midsize electric car emits 35,000 kilograms (77,161.8 pounds). Their electric car estimate appears high, but clearly demonstrates how much carbon is saved by riding an e-motorcycle instead of driving an electric car. It once again demonstrates that shaving a bit here and there doesn't make nearly as much a difference as a modal shift.

CAKE's exercise also demonstrates the complexity of the problem. There are as many parts in a motorcycle as there are in a building, and a lot less information. The shoemakers had an easier job of it.

The Future We Want: Supply vs Demand

When Elon Musk announced his new solar shingle a few years ago, he stood in front of a rendering of a house on a suburban lot with the solar shingle roof, a Tesla car and Powerwall battery in the garage, and a banner headline over the top: THE FUTURE WE WANT.

Musk is a "supply-side" kind of guy. He wants to make stuff and sell it to you, claiming this is the way we will get off fossil fuels.

More recently, MacArthur Prize winner, inventor, and entrepreneur Saul Griffith and the organization he founded, Rewire America, have been pitching "electrify everything," where we can have it all: "Same-sized homes. Same-sized cars. Same levels of comfort. Just electric." David Roberts of Vox is all in for this, calling it "the story that needs to be told about tackling climate change. Not a story of privation or giving things up. Not a story of economic decline or inexorable ecological doom. A story about a better, electrified future that is already on the way."[29]

Griffith and Roberts are also supply-side guys (they are almost always guys), envisioning a world of renewable power feeding our single-family suburban homes with solar panels on the roof and electric cars in the garage. The problem with almost all of the supply-sider solutions is that they don't scale; there is not enough lithium, copper, space, money or, most importantly, time. And of course, they have not even thought about the upfront carbon emissions of it all.

Then there are the demand-side guys, who believe that we have to stop consuming so many resources and have to reduce the demand for energy and fossil fuels. Engineer Nick Grant thinks they are two different personality types:

> My hunch is that the world is divided into "less is more" demand-side people, who focus on reducing their demand for emissions-heavy products and processes, and "less is a bore" supply-side people, who look for solutions within the sector that emits them—it is a personality trait.

Demand-siders dream of bicycling to small homes that keep warm from body heat. This is not Robert's "story of economic decline," but a positive vision that is achievable. But it's boring. Grant notes that a friend of his with a roof covered in solar panels brags about how much money he saves, but that there is no such excitement from sealing air leaks. Solar panels and batteries are expensive, and, to a certain extent, are what Thorstein Veblen might have called conspicuous consumption, while insulation and air sealing is invisible and inconspicuous. Grant wrote in his 2012 article for *Building Design*:

> I suggest that if we continue to prioritise a supply-side approach, we will be locked into a spiral of increased consumption. We can cover all surfaces with PV, cut down all the forests for biomass, plant palm oil to fuel our vehicles and install heat pumps in all buildings, but demand will always be just ahead of supply. Success should not be judged by the amount of green kit installed but by how close to sustainable levels of consumption we get.[30]

Grant wrote this a decade ago, before upfront carbon was on the radar, which significantly changed our understanding of sustainable levels of consumption. He concludes: "We need to rethink our entire approach and match demand to supply rather than fleeing into the future in a futile attempt to match supply to demand."

More recently, geologist Simon Michaux calculated the amount of material needed to make all the stuff required to electrify everything.[31] The numbers are vast; with lithium alone, "so enormous, it becomes now appropriate to ask is it even possible in context of mineral reserves available, as this far exceeds global reserves and is not practical." He also concludes that the only solution is to reduce demand.

> Replacing the existing fossil fuel powered system (oil, gas, and coal), using renewable technologies, such as solar panels or wind turbines, will not be possible for the entire global human population. There is simply just not enough time, nor resources to do this by the current target set by the world's most influential nations. What may be required, therefore, is a significant reduction of societal demand for all resources, of all kinds. This implies a very different social contract and a radically different system of governance to what is in place today.

Rethinking our approach means drastically reducing demand, which also requires a change in mindset, a change in the way we live. As Grant says, a change in our approach. As Michaux says, a different social contract.

Elon Musk sells cars, batteries, and solar roofs; Saul Griffith started companies that sell heat pumps and battery-boosted induction ranges. They believe that this is the future we want. It may well be, but it is an expensive future accessible and affordable for a small group of people with big rooftops. It's a future that very few people can have, and it's not the future we need.

Dematerialization and Degrowth

The most difficult and probably weakest chapter in my book *Living the 1.5 Degree Lifestyle* included a discussion of degrowth, defined by author Jason Hickel in his book *Less Is More: How Degrowth Will Save the World* as "a planned downscaling of energy and resource use to bring the economy back into balance with the living world in a safe, just and equitable way." He calls for "an economy that is organized around human flourishing instead of around capital accumulation; in other words, a post-capitalist economy. An economy that's fairer, more just, and more caring." Talking degrowth in North America is like touching the third rail; capital accumulation is its raison d'être. Hickel writes:

> Capitalism is fundamentally dependent on growth. If the economy doesn't grow it collapses into recession; debts pile up, people lose their jobs and homes, lives shatter. Governments have to scramble to keep industrial activity growing in a perpetual bid to stave off crisis. So we're trapped—growth is a structural imperative—an iron law.

That's why there is so much excitement about Green New Deal, The Future We Want, and Electrify Everything. Hickel writes:

Improvements will enable us to "decouple" GDP from ecological impact so we can continue growing the global economy forever without having to change anything about capitalism. And if this doesn't work, we can always rely on giant geo-engineering schemes to rescue us in a pinch. It's a comforting fantasy.

No wonder mentioning the word degrowth paints you as a raving commie.

If degrowth is too harsh a word, others, such as Samuel Alexander, have softened it by calling it a sufficiency economy:

> This would be a way of life based on modest material and energy needs but nevertheless rich in other dimensions—a life of frugal abundance. It is about creating an economy based on sufficiency, knowing how much is enough to live well, and discovering that enough is plenty.[32]

Emmanuel Macron, the president of France, preaches *sobriété*, or sobriety, which sounds a bit like degrowth and sufficiency. Indeed, Oliver Sidler of the négaWatt Association explains: "Sobriety means reducing our consumption by changing our lifestyles, uses and behaviour," and, "Sobriety is not synonymous with scarcity; it allows us to satisfy the same needs differently, by optimizing our consumption and preserving resources. It can certainly lead to a bit of degrowth, but not less happiness, because it makes our societies more resilient."[33]

Sufficiency and sobriety are presented here as lifestyle choices. However, degrowth is different; it is the inevitable result of sufficiency, of sobriety, of demand-side mitigation, of using less stuff because the world's economy is essentially about making stuff, whether it is green or not. Economist and physicist Robert Ayres noted, "The economic system is essentially a system for extracting, processing and transforming energy as resources into energy embodied in products and services," or, as I simplified it in my last book, the purpose of the economy is to turn energy into stuff. Ayres notes that there is useful energy, called exergy, that you get out of fossil fuels as heat, which can run engines. There's lots of low-grade energy around, like the warmth in the ocean or the air around us, but it doesn't do much work.

Ayres doesn't have much time for traditional economics, but teaches that economics is subject to the laws of thermo-dynamics, which is very convenient when thinking of energy and carbon.

> The first law of thermodynamics, conservation of mass/ energy, says (among other things) that all industrial processes—extraction, reduction, synthesis, shaping and forming—generate waste residuals. The mass of residuals from industrial activity far exceeds the mass of materials embodied in "final products," all of which eventually also become wastes. This means that "zero waste" is a far distant goal that can never be achieved in the real world.

The second law of thermodynamics has even bigger impli-cations.

> The second law, known as the entropy law, says that every process in the universe converts order (low entropy) into disorder (high entropy). This means that economic processes that seem to produce local order (think pat-terns) do so only by increasing global disorder. That, in turn, means that all economic processes generating useful goods and services can only do so by consuming— literally destroying—exergy.[34]

Those economic processes consume exergy and turn it into goods and services, waste heat, upfront carbon emissions, and eventually landfill. So those of us who call for fixing things instead of buying new, of buying less, of using less stuff on our buildings and our lives, are not doing our main economic func-tion: consuming energy in the form of goods and services. As prominent degrowther Giorgos Kallis notes in an essay, "Radical Dematerialization and Degrowth," renovating and retrofitting buildings use more labour and fewer materials than knocking buildings down and replacing them, but produces less profit. A retrofit economy is possible, but its houses and cars may look more like Cuba's than California's."[35] Investment in renewable energy creates jobs, but then those solar panels just sit there generating power; whereas, with fossil fuels, you always have to dig up more stuff.

> Third, consider a scenario where people fix their own phones or their cars instead of buying new ones. The circulation of new goods in the economy would slow down. The unpaid, domestic portion of the economy will grow, but market economy will shrink. The surpluses that drive growth will diminish, and therefore growth will decline. An economy that grows 2% per year has to double the speed with which it churns goods and services every 35 years. There must always be demand to satisfy the bigger and faster production. Production requires consumption. Planned obsolescence and an ever-faster turnover of goods are necessary if growth rates of 2–3% per year are to be sustained.

A world where we must use less stuff to reduce upfront carbon emissions and make less stuff with operating emissions means that economic degrowth isn't a choice, it's an inevitable result. But it doesn't have to be an uncontrolled or random shrinking of the economy, causing what might be called a recession or depression. As author Stan Cox notes:

> Degrowth differs fundamentally from a recession. A recession is a reduction in GDP, one that happens accidentally, often with undesirable social outcomes like unemployment, austerity, and poverty. Degrowth, on the other hand, is a planned, selective and equitable downscaling of economic activities.... Associating degrowth with a recession just because the two involve a reduction

of GDP is absurd; it would be like arguing that an amputation and a diet are the very same thing just because they both lead to weight loss.[36]

Kallis's suggestions for planned and equitable policies to deal with the problem are unsatisfactory. He suggests work-sharing, to spread the jobs around and redistribute the gains from productivity. Taxation should shift from being based on income from work, which is a "good" and instead tax resources and carbon, which are a "bad." Oh, and the whole banking system has to be changed because "currently private banks create money and lend it with interest. Repaying interests creates an imperative for the economy to grow." And capitalism!

> Understandably, some prefer to imagine, against historical experience, that it is possible to decouple resource use from economic growth, and to fuel growth with renewable energy. The technological achievements during the last 200 years or so of capitalism fuel this hope. Yet previous technological achievements were energy and resource-intensive, accelerating the transformation of matter and the fresh occupation of new territories. Capitalism has been extremely good at relentlessly and violently expanding; there are few signs that it can be as good at peacefully contracting.

Kallis concludes, noting it might well be that "the politically acceptable is ecologically disastrous while the ecologically necessary is politically impossible." But you can't avoid the inconvenient truth: "further growth is not ecologically sustainable."

Or is it? Can we have some growth, carefully managed to get more exergy out of our energy and reduce waste heat and carbon emissions? Can we have frugal abundance, sobriety, and sufficiency and still have much of an economy? If degrowth is inevitable, can we have what Samuel Alexander calls a "prosperous descent"? Can we enjoy the ride?

Enjoy the Ride with Demand-Side Mitigation

Demand-side mitigation is "living the 1.5-degree lifestyle"—making changes that mitigate your carbon emissions through your actions and how you live. This is a rich world's problem, as was my living the 1.5-degree lifestyle; most people in the world have nothing to mitigate. But as the richest 10 percent of the global population (which, according to OXFAM, includes most North Americans and Europeans) emit 50 percent of the carbon, we are a good place to start.[37] Several studies published since I wrote my book confirm its findings and come up with a level of energy consumption consistent with well-being.

The first study was published a few months before the IPCC Working Group III report and has many of the same authors, and much of it is in the report. It discusses how we can reduce carbon emissions and still live a good life, and is titled "Demand-Side Solutions to Climate Change Mitigation Consistent with High Levels of Well-being."[38]

Lead authors Felix Creutzig and Leila Niamir write that demand-side mitigation strategies in the building, transport, food, and industry sectors could reduce emissions between 40 and 80 percent, depending on the sector. There are three types of strategies employed:

- "Improve" options include more efficient building envelopes, appliances, and more efficient energy use by industry sectors.
- "Shift" options are related to transportation, including a modal shift to walking, cycling, and shared mobility. It also applies to food, shifting to flexitarian, vegetarian, or vegan diets. "These are options that require physical and choice infrastructures that support low-carbon choices, such as safe and convenient transit corridors and desirable and affordable meat-free menu options," write the authors. "They also require end users to adopt these choices, individually and socially."

- "Avoid" options are across the board. "Cities play an additional role, as more compact designs and higher accessibility reduce demand for distanced travel and car mobility and also translate into lower average floor size and corresponding heating, cooling and lighting demand," the authors write.

In the previous chapter, I questioned whether we could have a "prosperous descent" with "frugal abundance"; this study suggests that we can.

> Our study shows that, among all demand-side option effects on well-being, 79% (242 out of 306) are positive, 18% (56 out of 306) are neutral (or not relevant/specify) and 3% (8 out of 306) are negative. Active mobility (cycling and walking), efficient buildings, and prosumer choices of renewable technologies have the most encompassing beneficial effects on well-being with no negative outcome detected.

Not only did demand-side mitigation result in lower carbon emissions, but, surprise! People were healthier, the air and water quality was better, there was less waste, and people used less stuff. There was greater social interaction and a greater general level of happiness.

More recently, a Stanford University study looked at human well-being and per capita energy use, which found that at a certain point, using more energy doesn't make you happier.[39] The problem with our history with energy is that access to it has made our modern civilization possible, but as Max Roser of Our World in Data notes, it is not evenly distributed. "The first energy problem of the world is the problem of energy poverty—those that do not have sufficient access to modern energy sources suffer poor living conditions as a result." But it appears you can have too much of a good thing.

The study looked at nine measures of health and well-being, including health, economic and environmental well-being, income inequality, happiness, infant mortality, life expectancy, prosperity, and sanitation. The researchers plotted these against the average energy consumption for countries worldwide.

The global average per capita energy consumption is 79 gigajoules; Americans average 284. But the study found that above 75 gigajoules, adding energy didn't add much to health or well-being; gigajoules don't buy happiness. Study co-author and climate scientist Anders Ahlström noted, "Energy supply is similar to income in that way: Excess energy supply has marginal returns."

Vaclav Smil said much the same thing in his earlier book *Energy and Civilization*:

> Satisfying basic human needs obviously requires a moderate level of energy inputs, but international comparisons clearly show that further quality-of-life gains level off with rising energy consumption. Societies focusing more on human welfare than on frivolous consumption can achieve a higher quality of life while consuming a fraction of the fuels and electricity used by more wasteful nations.

A recent British Study, *Social Outcomes of Energy Use in the United Kingdom: Household Energy Footprints and Their Links to Well-Being*, put the happy gigajoule number at 100, a bit higher than the Stanford study, but still a fraction of what the not-much-happier rich jet-setters who emit more carbon just from transport than the 100-gigajoule user does just living.[40] One critic noted that "Car journeys and flights taken by the richest British people—especially 'white, wealthy middle-aged men'—used more energy that year than 60% of the population got through in total."

But, when the researchers correlated a well-being score (WBS)—a mix of mental and physical health, loneliness, financial security, and housing, to energy consumption—they found that being rich above a certain point did not automatically result in a high WBS. There are evidently a lot of miserable high-fliers and also a lot of happy middle-income people with low energy consumption and carbon emissions. The study authors conclude:

> Without policies aiming for sufficiency, we will not be able to mitigate the effects of our lifestyles. Living a sufficient lifestyle does not doom us to "go back to caves." Our analysis suggests that more efficient energy services, such as the provision of public transport and improvements in housing, could substantially lower energy demand without adversely affecting well-being outcomes. However, this will not be enough, interventions must also target high-energy users whose energy excess can undermine efforts to reduce energy consumption. Sufficiency can mean flourishing for all, but sustaining the status quo of unchecked energy-intensive lifestyles of a few rich can be also disastrous for all.

So, we now have the IPCC Working Group III mitigation report, a pile of studies, proponents of degrowth, demand-side mitigation, living the 1.5-degree lifestyle, sobriety and sufficiency all calling for the same thing: a different way of living, with lower energy consumption and lower carbon emissions. But what are the strategies that we can use? What will it look like?

Why Sufficiency Is the Solution

We are living in marvelous times. Wind turbines and solar panels have made renewable electricity cheaper than any other kind. We can heat and cool with heat pumps and cook with magnetism and we really don't have to burn stuff anymore. With Passivhaus designs becoming more common and affordable, we won't need much heating or cooling in the first place. Electric cars are dropping in price and will soon make gasoline-powered cars functionally obsolete. Everything is becoming so damned efficient. But as Samuel Alexander noted in his Critique of Techno-Optimism, efficiency without sufficiency is lost.

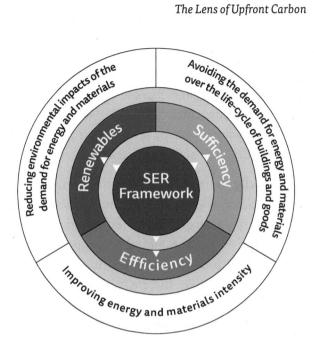

SER Framework. Credit: Yamina Saheb

In order to take advantage of efficiency gains—that is, in order for efficiency gains to actually *reduce* resource and energy consumption to sustainable levels—what is needed is an economics of sufficiency; an economics that directs efficiency gains into reducing ecological impacts rather than increasing material growth.[41]

Ecological limits mean more than just carbon. We know how to make aluminum without carbon emissions, but we are still digging up bauxite by the boatload. We can build towers out of wood, but we also need forests for biodiversity. So even as what we build gets more efficient and the upfront carbon emissions of our materials are reduced, we still need to use less of them.

Many have been writing about the need for increasing energy efficiency and switching to renewable resources, but Yamina Saheb of Lausanne University, one of the lead authors of the IPCC Working Group III report on mitigation, added sufficiency in her Sufficiency/Efficiency/ Renewables (SER) Framework. She defines it:

> Sufficiency is a set of policy measures and daily practices which avoid the demand for energy, materials, land, water, and other natural resources, while delivering well-being for all within planetary boundaries. Sufficiency bridges the inequality gap by setting clear consumption limits to ensure a fair access to space and resources.[42]

Saheb is primarily addressing buildings, but also notes that the concept of sufficiency goes well beyond that, setting a lower limit of sufficiency that provides a decent standard of living, which is "a set of essential material preconditions for human wellbeing which includes housing, nutrition, basic amenities, health care, transportation, information, education, and public space."

The framework is somewhat confusing because "efficiency" has two subcategories, basic operating energy efficiency and what she calls "materials intensity," which is what we have called design efficiency. Similarly, "renewables" includes both energy and materials. But the point is significant: it is a three-legged stool. Remove any one sufficiency, efficiency, or renewable and the concept fails.

With buildings, Saheb calls for optimizing their use, repurposing existing buildings, prioritizing multifamily dwellings, and "adjusting the size of buildings to the evolving needs of households by downsizing dwellings." This might be done with progressive taxes based on per-capita floor area.

But housing doesn't exist in a vacuum. What we build is a function of how we get around, or as transportation planner Jarrett Walker noted, transportation and land use are the same things spoken in two different languages. How we live, shop, work, and get between everything must be examined through the lens of sufficiency.

Samuel Alexander took a shot at this in another essay from 2012, "The Sufficiency Economy: Envisioning a Prosperous Descent," in which he suggests how we might live in a world where everyone has enough but no more than they need. He defines it:

> The fundamental aim of a sufficiency economy, as I define it, is to create an economy that provides "enough, for everyone, forever." In other words, economies should seek to universalize a material standard of living that is sufficient for a good life but which is ecologically sustainable into the deep future.

It is a utopian vision—some might call it dystopian—and it is not without problems, but it is a good place to start. It provides a good list of the different aspects of our lives and how they would adapt to a sufficiency economy. He has an interesting take that is worth considering as a starting point.

Alexander looks at all aspects of our lives:

Water: Use less of it in the developed world. "Enough to live a dignified existence without leaving much room for waste." That would mean shorter showers and washing our clothes less frequently, and the widespread use of composting toilets.

Food: Much of it is grown with fertilizers made from natural gas, pesticides made from oil, wrapped in plastic made from oil and shipped long distances in trucks powered by oil. In a sufficiency economy, Alexander says we should all eat local organic food fertilized by compost from our toilets. We would have vegetables in our front yards, chickens in the rear, and herbs on our balconies.

Clothing: Alexander writes that "the consumption of clothing, like all consumption, is a culturally relative social practice, so as more people came to wear second-hand or sustainably designed clothing, new social standards would be quickly established."

Work and Production: Alexander doesn't draw a particularly happy picture, anticipating tough economic times and high unemployment levels. But we will be busy: "The provision of basic needs—such as food, clothing, shelter, tools, and medicine—would be the primary focus of production, and the motivation would be to produce what was necessary and sufficient for a good life, rather than to produce luxuries or superfluous abundance sustenance."

Housing: Alexander writes that we will not be living in eco-houses and earth ships, but in our existing suburbs, which will need to be retrofitted and intensified. He also suggests filling our backyards with shipping containers and wigwams, being small and cheap accessory units.

Transportation: Here, Alexander is prescient, saying much the same thing I have been saying in this era of EV-mania, or that Simon Michaux has concluded: "In order to decarbonise the economy, it is required that people drive much less, or not at all. Electric cars will not be able to escape this imperative, because producing them depends on fossil fuels."

Energy: Alexander believes we will need less energy in a sufficiency economy, but more renewables. But he worries, like Michaux, about the resources needed to build them all.

Technology: Here, he sounds like me talking sufficiency a decade ago: "The clothesline will replace the clothes dryer; the bike will largely replace the car; and the television will essentially disappear."

This is Alexander's picture of a sufficiency economy. I wish he wouldn't call it "a prosperous descent" because descent somehow sounds negative. This is a positive picture of healthy, resilient living where everyone has enough of what they need. I would say that things are looking up.

Unlike operating carbon emissions, where efficiency rules, dealing with upfront carbon emissions is all about sufficiency, about using less of everything, because everything has a footprint. In the next section, we will look at specific strategies for getting by with less. Then we might develop our own vision of a prosperous ascent.

CHAPTER 2

STRATEGIES FOR SUFFICIENCY

I am an architect, and much of my thinking and writing started with buildings, but I have long worried about consumption issues. When asked to write a bio for Treehugger, I described my thinking twenty years ago: "He has become convinced that we just use too much of everything—too much space, too much land, too much food, too much fuel, too much money—and that the key to sustainability is to simply use less." For my bio in my book, I added that this is "what he calls radical sufficiency." I would add now that sufficiency is the best, and perhaps the only, way to deal with the problem of upfront carbon emissions: how much of this stuff do we need? How can we design our lives and our world to use less?

In terms of Maslow's hierarchy of needs, the function of a building is first to meet our basic physiological needs of food, water, shelter, and warmth. But a building alone cannot do that; it is part of a larger built environment that delivers clean water and takes away poopy water, that provides the fuel for warmth, and access to other buildings that supply the clothing and food needed for warmth and sustenance. The designs of our homes and buildings are predicated on the way we get between them; your typical suburban family can shop at the Walmart superstore because they have the SUV to carry everything back to their home, which is big enough to store it all.

All of this has been possible because of what can truly be called embodied energy stored in fossil fuels. They power our food system with transport, fertilizers, and the cold chain; our built environment of buildings and the vehicles that connect them; the plastics that make our linear economy work.

As noted previously, most of the carbon dioxide we have added to the atmosphere has come from operating emissions, moving vehicles, generating power, heating and cooling our homes and buildings, or making materials such as steel, concrete, and plastic to make things. People are working madly to reduce those operating carbon emissions, which is why we got our ironclad rule of carbon. To reiterate:

> As our buildings and everything we make become more efficient and we decarbonize the electricity supply, emissions from embodied and upfront carbon will increasingly dominate and approach 100% of emissions.

The main methods of reducing and eliminating operating emissions are relatively straightforward: increase efficiency and switch to zero-carbon energy sources. Upfront carbon emissions are not so obvious or simple. There are several universal approaches and strategies for efficiency, including materiality, simplicity, frugality, ephemerality, and circularity. But most importantly, I keep coming around to sufficiency, how much we need.

Materiality: Build Out of Sunshine

Materiality, the stuff that things are made of, has been applied mostly to buildings, but the question can be asked of any object or article any time there is a question of whether one uses natural materials that were made with sunshine, or those that are manufactured or mined.

When I reviewed Bruce King's book *The New Carbon Architecture*, I complained about his subhead "Building Out of Sky," thinking it didn't have a ring to it and suggested "Building Out of Sunshine" instead. But King was right; sunshine is the engine driving the chemical reaction that takes carbon out of the carbon dioxide in the air and water out of the rain that falls to give us cellulose and lignin and everything that plants are made of, with oxygen as the "waste" carelessly vented back into the atmosphere.

Once upfront carbon became an issue, the building industry started looking for strategies to deal with it. The first and most obvious one is to use materials with less upfront carbon instead of traditional concrete and steel; this is why we have seen the rapid proliferation of mass timber and other materials grown instead of mined.

There is nothing new about mass timber; we have been building with it for thousands of years. Multi-story wood warehouses, factories, and apartment buildings were built across North America. Then, after the great Chicago fire, architects started building with iron and steel, while the big trees needed for the columns and beams were getting harder to find and more expensive. Building codes were changed to ban wood construction in taller buildings, ostensibly for fire safety.

Europe has always had a sophisticated forestry industry, and in the 1990s, the landlocked Austrians had more expensive wood than they could use. In 1996, an academic research program came up with the idea of Cross-Laminated Timber (CLT) as we know it today, as a way of adding value to Austrian lumber—much more on this subject later in the book. Architects are eating it up, seeing it as a great way to reduce their upfront carbon emissions.

Years ago, I wrote about the Enterprise Centre in the UK, describing it as the future of green building. Its exterior was clad in traditional thatch; materials used inside included hemp, wool, nettles, reeds, and recycled wood. I joked that you could chop up this building, add milk, and enjoy a high-fiber breakfast; I described it as almost edible.

That may be a stretch; edibility is used tongue in cheek, but one way to eliminate the upfront carbon of the materials from which we make things is to grow them instead of digging for them. Mushroom mycelium is mixed with flax, canola, and hemp to make acoustic and thermal insulation. Cork is used for insulation, bonded to cotton and used as fabric. Tree bark has been turned into cloth. Fish skins have been turned into leather. Peanut shells have been turned into particle board.

Bamboo is used as a structural material, but also turned into rayon-like fabrics. While the traditional rayon process uses toxic chemicals, the Lyocell process is nontoxic and uses far less water; eucalyptus wood is put through the Lyocell process and is turned into running shoes. Nylon substitutes are now being made with castor oil, PET substitutes such as PLA are being made with cornstarch, and polyethylene can be made from ethanol, which is made from corn.

Perhaps the natural material with the brightest future is straw. Straw bale construction has been the preserve of hippies for decades, as has cob, where straw is mixed with clay. However, now straw is used as insulation in prefabricated panels, mixing some of the newest and oldest technologies.

But sometimes, this is all more greenwash than green. PLA bottles are used as an excuse to continue using single-use plastics and call them biodegradable, except they don't biodegrade well when mixed in the regular waste stream. They can contaminate it all. It's made from corn and, as with ethanol used for fuel, raises questions of whether we should be feeding cars and bottle makers instead of people. Cheap bamboo fabrics made in the traditional viscose process are a major source of pollutants and the clothes are made in the same terrible sweatshops.

Making any product has some degree of upfront carbon emissions. A few years ago, I thought "Agriboard," a particle board made of waste straw, was the new green wonder; it turns out it was made with heat from natural gas. Even our beloved mass timber requires trucks, kilns, and glues, and chews up four times as many trees as traditional wood frame construction, impacting forests and biodiversity. The taller the building gets, the more fiber it requires.

Even when we are making things out of sky, out of carbon dioxide and water, there are limits. Even with the greenest of materials, sufficiency still matters.

Materiality: Use Less Stuff

Will Arnold of the Institution of Structural Engineers writes:

> Most approaches to reducing structural emissions fall into one of two types of action. You can minimise the amount of material that you use (put simply: use less stuff), or you can minimise the amount of carbon released when producing those materials. These also form the two parts of the equation describing embodied carbon emissions:
>
> Embodied carbon = (quantity of material) × (carbon intensity factor)
>
> As a building designer, your priority is the first type of action or the left-hand side of the equation. You must push to use less stuff. This involves prioritizing better use of existing building stock and then configuring new structures to minimize material use. Our massing, layout and configurations must get more efficient (we often need to convince others to enable this), and then our design methods and utilisations must deliver this with no "spare fat."

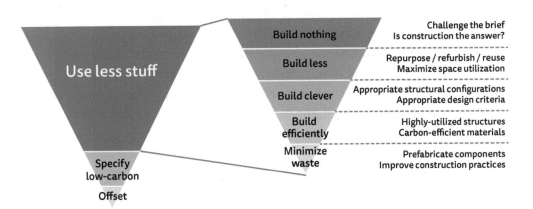

Use less stuff. Credit: Will Arnold

This is not just a hippie academic preaching to the wilderness; there are a lot of people saying the same thing, people who are in the business of designing stuff that takes a lot of stuff to make. "Stuff" seems to be the word of the year. Others are more academic in their approach and use fancier words than "stuff." Engineering giant Arup writes:

> This strategy aims at meeting the project requirements with minimal material consumption. At all levels, it fosters simple design approaches, thoughtfully considering the real need of components and materials. It aims at questioning if certain components can be refused without compromising the ability for the project to function at the desired performance level.[1]

And when you look at that inverted pyramid, building out of wood shrinks in significance, way down below using less.

And the real importance of Arnold's pyramids and his formula is that they don't just apply to buildings, they apply to everything, whether it is a car, a computer, or an article of clothing. You want to use the minimum amount of stuff, and the stuff you use should have the least amount of upfront carbon.

Materiality: Use Less of the Bad Stuff

A common response to my complaints about the upfront carbon in everything is that steel and aluminum are getting greener and better and will soon have much lower upfront emissions than they do now. Sure, currently, 8 percent of global CO_2 emissions come from steel production and 2 percent from aluminum, but technology! And recycling! But let's have a look at the reality of this.

Steel

Steel is indeed recycled in huge quantities. In North America, electric arc furnaces (EAF) dominate the industry and reduce the carbon emissions from making a tonne of steel between 50 percent and 70 percent, compared to blast furnaces, depending on the cleanliness of the electricity. According to the Carbon Smart Materials Palette from Architecture 2030, "Using steel from electric arc furnaces is the best way to reduce embodied emissions in steel because EAFs use high levels of recycled material and can be powered by renewable energy sources."[2]

New technologies are also revolutionizing virgin steel production. Currently, it takes over three-quarters of a tonne of metallurgical coal to make a tonne of steel in a Basic Oxygen Furnace (BOF) in a reaction where the coal is cooked to make coke, which is mixed with the molten iron ore. The carbon reacts with the oxygen in the ore to make carbon dioxide, leaving pure iron behind. However, hydrogen can also be used to bond with oxygen to make water vapor, and the hydrogen can be made with electricity, resulting in steel with very low carbon emissions.

While there are reasons for optimism, the problem is that the steel demand is always growing, so there is never enough recycled steel. A study looking at global steel demand found:

> Steel production using recycled materials has a continuous growth and is likely to be a major route for steel production in the long run. However, as the global average of in-use steel stock increases up to the current average stock of the industrialised economies, global steel demand keeps growing and stagnates only after 2050. Due to high steel demand levels and scarcity of scrap, more than 50% of the steel production in 2050 will still have to come from virgin materials.[3]

According to a Wood Mackenzie study, converting the virgin steel industry to hydrogen would require 52 million tonnes of green hydrogen by 2050, equivalent to about three-quarters of all the hydrogen produced today, almost none of which is green. Making it would require being powered by 2,000 gigawatts of renewable electricity, "equivalent to two-thirds of today's global clean-energy generation and more than double that of the world's current installed wind and solar capacity."[4]

These numbers are mind-bogglingly huge. Furthermore, in 2010, over half of the world's steel was made in China, which is not likely to be made with wind-powered green hydrogen any time soon. China is also the world's largest exporter of steel, both as raw material and in finished products.

This is why we can't just say don't worry, steel is getting greener, or engineers are learning how to use it more efficiently. Fifty-one percent of steel goes into buildings and infrastructure, so we have to use less of it by just building less and using low-carbon materials. Twelve percent of it goes into cars and trucks; it wouldn't hurt to have fewer of those.

Aluminum

Aluminum is touted as the most wonderful of metals; it is one of the most common on Earth. It's the most valuable item in the blue bin, with about 65 percent of household aluminum recycled. But only 17 percent of aluminum is used in packaging; 26 percent goes into buildings, and 36 percent into transportation.[5]

The industry calls it a "circular economy solution; high value and infinitely recyclable aluminum is a material tailor-made for a more circular and sustainable economy." They claim "a recycled aluminum beverage can, car door or window frame is often recycled directly back into itself. And this process can happen virtually infinitely."[6]

This is not quite true; many products, such as Apple computers and airplanes, cannot use recycled aluminum; they need virgin aluminum to get the alloys just right. If you care about carbon emissions, different sources of aluminum can be all over the map, ranging from half a tonne of carbon emitted per tonne of aluminum to 18 tonnes. And even if they say it is recycled, it could still be greenwashed.

The nascent hydrogen industry has developed a color label system, which I am going to imitate to categorize aluminum by its carbon emissions. But first, a little background about how aluminum is made.

The making of aluminum starts with bauxite, a sedimentary rock that is strip-mined in Jamaica, Guinea, Russia, Malaysia, Australia, and Brazil. It is an environmentally destructive process. In his book *Aluminum Upcycled: Sustainable Design in Historical Perspective*, Carl Zimring writes, "As designers create attractive goods from aluminum, bauxite mines across the planet intensify their extraction of ore at lasting cost to the people, plants, animals, air, land and water of the local areas."

Bauxite is about half alumina, which is separated from "red mud," which is responsible for most of the environmental damage. It also has a cost in carbon; according to the Financial Review, it takes about 2.5 megawatt-hours of electricity to make one tonne of alumina, and a lot of the world's best refineries draw that power from gas generators. At 185 kilograms (407.9 pounds) of CO_2 per mWh of gas-fired electricity, that's close to half a tonne of CO_2 per tonne of alumina. Since it takes two tonnes of alumina to get a tonne of aluminum, that's about a tonne of CO_2 per tonne of aluminum.

Alumina is aluminum oxide; the hard part is separating the aluminum from the oxygen. Most aluminum is made with the Hall-Héroult process: you dissolve alumina in cryolite or synthetic sodium aluminum fluoride and run a huge electric charge through it; it takes about 15 kilowatt-hours per kilogram of aluminum. The oxygen in the alumina combines with carbon from the anode to make about four tonnes of CO_2 per tonne of aluminum.

This is why aluminum has been nicknamed "solid electricity," and why most of my proposed colors are based on the source of the electricity.

Brown Aluminum

Brown aluminum is made with coal-fired electricity, as is done in China, Australia, and the United States, with a carbon footprint of about 18 tonnes of CO_2 per tonne of aluminum. China has 56 percent of the world aluminum market.

Gray Aluminum

In Saudi Arabia, the aluminum smelters run on electricity from natural gas, part of the kingdom's Vision 2030, which is a plan to diversify its economy and make it less reliant on the ever more volatile global petroleum trade. It has a carbon footprint of about eight tonnes of CO_2 per tonne of aluminum.

Light Blue Aluminum

The industry likes to call aluminum made with hydroelectricity green, but I like blue because that's the color of water, and it is not carbon free if it is made with the Hall-Héroult process, which gives it a footprint of about four tonnes of CO_2 per tonne of aluminum, plus a tonne from the alumina.

Dark Blue Aluminum

This is the wonderful new process developed by ELYSIS and funded by Apple and the Canadian government, where the carbon anode is replaced with "a ceramic anode for aluminum production that emits only oxygen and no greenhouse gases and last 30 times longer than those made from conventional materials."[7] Others are now using "inert anodes" in Scandinavia and Russia. But I still don't think it can be called carbon free; we still have that tonne of CO_2 from the alumina production.

Light Green Aluminum

The crowd went wild when Apple announced that the new MacBook Air (that I am writing this on) was made with 100 percent recycled aluminum. But Apple needs precise alloys, so they are using what is sometimes called "pre-consumer waste" or "process scrap." Having lots of pre-consumer waste means that you are probably doing something "wrong" in your manufacturing processes; it is not a badge of honor.

Ford does the same thing, pleading virtue while picking up about 25 percent of the aluminum that is left after hoods, tailgates, doors, and fenders are stamped into shape. "The closed-loop recycling system has helped Ford recycle more than 20 million pounds of aluminum a month, which is enough to build more than 37,000 F-series truck bodies monthly." But nobody throws this away; this is not recycling.

Some say it is even worse. Norsk Hydro point out that melting and processing the process scrap emits about 0.5 tonnes of CO_2.

> However, the process scrap has never fulfilled its purpose as a product and thus carries the carbon footprint of the original primary aluminum from which it is produced. This means that the carbon footprint of recycled process scrap is not 0.5 tons CO_2 per ton of aluminum but the carbon footprint of the original primary aluminum PLUS 0.5 tons CO_2 per ton of aluminum. Some companies claim that the carbon footprint of process scrap is equal to the carbon footprint of post-consumer scrap. Such an approach would then equalize hydropower-based aluminum and coal-based aluminum as soon as the metal is processed. This approach would lead to "greenwashing" and would favour industrial inefficiencies.

This will be controversial, but I believe they have a point, so my color chart has 0.5–18.5 tonnes.

Dark Green Aluminum

Finally, we have the only truly green aluminum, which is recycled from post-consumer waste. This is where we truly have to go, to a closed loop where we stop the hugely destructive mining of bauxite and processing it into alumina. The recycling rate of aluminum is high at 67 percent, but the rate for packaging is far lower at 37 percent. Much of that goes into foil pouches and multilayer materials, such as Tetra Paks, that can't be recycled affordably. That's why we must design for deconstruction and disassembly so that materials can be recovered easily and avoid what Bill McDonough called "monstrous hybrids" that can't be taken apart.

And even this is not zero carbon; just processing it probably generates half a tonne of CO_2 per tonne of aluminum. This is why, ultimately, we need to switch back to refillable containers. We have to stop creating demand for new aluminum, because as Carl Zimring noted, "even such intense and virtuous recycling that we do with aluminum, even if we catch every single can and aluminum foil container, it's not enough. We still have to use less of the stuff if we are going to stop the environmental destruction and pollution that making virgin aluminum causes."

Plastic

I bought my wife a new teakettle for Christmas. Her old one worked perfectly well, except it was made of plastic, and I had recently read Matt Simon's book *A Poison Like No Other*, which discussed how plastics are getting into everything, including our bodies.

> Takeout coffee cups, which are lined with polyethylene to keep the paper from disintegrating, release tens of thousands of microplastics and millions of nanoplastics—along with nasties like lead, arsenic, and chromium—into hot water in just 15 minutes, one experiment found. Teabags are also made of plastic, and they shed particles when steeping. Electric kettles inject tens of millions microplastics into a liter of water.

Fortunately, Kelly doesn't use teabags and now has a fancy stainless steel kettle that should last for a while. We think of plastic as long-lasting, but reading Simon's book makes you realize that everything made of plastic is constantly shedding micro and nanoplastics. They will have micro and nano carbon footprints, but it will add up. According to the United Nations Environment Program (UNEP), more than 400 million tonnes of plastic are produced yearly; 36 percent, or 144 million tonnes of that is single-use plastic designed for immediate disposal. The embodied or upfront carbon in plastic averages 3.31 kilograms (7.3 pounds) of CO_2e per kilogram, so single-use plastic generates 476 million tonnes of greenhouse gases yearly, more when you consider that much of it is incinerated. Single-use plastics alone emit half as much as aviation, but don't get half the attention. Textiles are 14 percent of plastic production, much of it fast fashion that is slightly more than single use. Building and construction are 16 percent, much of it insulation. We usually think of plastics as a pollution problem, but as Simon notes, "Plastics are fossil fuels, and plastics are climate change, so in scorning the material we'll tackle both crises—really, we can't fix one without fixing the other."

Erik Solheim, head of UNEP, introduced their report with the statement, "Plastic is a miracle material. Thanks to plastics, countless lives have been saved in the health sector, the growth of clean energy from wind turbines and solar panels has been greatly facilitated, and safe food storage has been revolutionized." Solheim says, "Plastic isn't the problem. It's what we do with it." I disagree. Plastic is the problem. We are making way too much of it, and it is ending up in all the wrong places.

Concrete

Concrete is our biggest problem; we use a lot of it. The global concrete industry has issued a road map claiming that they will be net zero emissions by 2050, but there is a lot of wishful thinking. Even they describe the reaction in the kiln that turns limestone into cement, releasing carbon dioxide, as the "chemical fact of life." The only thing they can do about it is try to make it disappear with carbon capture and storage.

Others are more optimistic. There are some exciting new ways of making the calcium oxide or lime needed to make cement. At Cambridge University, scientists are making Cambridge Electric Cement (CEC)—calcium oxide from recycled concrete as a by-product in electric arc furnaces. Sublime Systems is using electrolysis. Perhaps the most interesting idea comes from Biomason, who are replicating the natural process used to make the calcium carbonate seashells that were later squished into limestone: they are using bacteria to take calcium and CO_2 out of water and convert it to calcium carbonate.

The problem is scaling it up. We use so much concrete, and 56 percent of it is produced in China. Some think demand will be reduced; Dr. Cyrille Dunant of the Cambridge University team told me that "the slowdown in population growth implies we'll need only about 50–60 percent of today's needs. So in effect, blending [recycled concrete] which doubles the amount of product, plus population which halves demand, plus material efficiency, which halves it again suggests CEC could cover all future cement needs in 2050 with a margin."

That sounds optimistic, and a bit late. The only real answer, once again, is to use less of the stuff. This is doable to a certain extent. William McDonough + Partners recently completed one of the largest mass timber projects in North America, substituting CLT for concrete. But they could only reduce the amount of concrete in the building by a little more than a third; the rest was in the parking structure they were forced to build. Parking is what I have called the upfront carbon iceberg, unseen but deadly—another good reason to have fewer cars.

It's clear that for just about whatever material we use to make stuff, we have to use less, primarily because of the upfront carbon emissions. But there are several considerations and approaches that designers can take to reduce the amount of stuff we need or whether we need it at all. Some of these ideas go back to Aristotle and Diogenes, but let's start with Bucky Fuller.

Ephemerality

A decade ago, after selling Treehugger, the website he founded, Graham Hill started a business designing small apartments with moving walls and Murphy beds. I was shocked to find no stove in the kitchen; instead, Graham pulled a portable induction cooktop out of a drawer. If he was having a party, there were two more; he would take out what was needed when needed. His kitchen stove had been ephemeralized.

Ephemeralization is a term invented by R. Buckminster

Fuller in his 1938 book *Nine Chains to the Moon* to describe how, through technological advancement, we can do "more and more with less and less until eventually, you can do everything with nothing." He concluded:

Efficiency = doing more with less.

This was the thinking behind both the form and the structure of his geodesic dome; a sphere encloses the most volume relative to surface area, and the tensegrity structure that Fuller invented used the least amount of material to enclose the sphere.

We are all living in a world of ephemeralization, the best example being that smartphone in your pocket; thirty years ago, it would have taken an entire Radio Shack store's worth of equipment to do what it can do, which is why we no longer have Radio Shacks.[8] The iPhone is a marvel of miniaturization and energy efficiency, the very definition of ephemeralization.

Ephemeralization is a form of sufficiency: How do you do more with less and do it better? Let's look at that kitchen stove. In the mid-1800s it revolutionized cooking, as the box made from six plates of iron allowed multiple dishes to be cooked at once, heated the home, and saved women's lives as "hearth death" was the second-biggest killer of women after childbirth.[9] When gas and electricity replaced wood, the stove remained a hot box with the cooking surface on top of the oven, even though they were powered separately, and even though it was obvious that bending down to get stuff out of the oven was not convenient nor safe.

Now we are going through another cooking revolution, where the oven and the cooktop don't even use the same technology, and there is no good reason to put them together, unless, as in our house, you are replacing an existing appliance and don't want to renovate the kitchen. Then you find that putting them together is a marriage of inconvenience, where the oven's heat rises and heats the induction section and will probably lead to its eventual demise. Meanwhile, companies such as Italy's Fabita are selling induction hobs you hang on the wall and take down as needed. Others are selling tiny two-burner cooktops where people keep spare portable units for the occasional times they need more. What was once the dominant feature in the kitchen has been ephemeralized.

Ephemeralization may well solve a major problem we face as we electrify everything: what author Richard Heinberg called "peak everything," where we are running up against limits to natural resources. Copper is a good example of this. Where a gasoline-powered car has about 23 kilograms (51 pounds) of copper in its wiring and motors, an electric vehicle has about 83 kilograms (183 pounds). According to Bloomberg:

> There is a challenge facing this growth trajectory, and it's not so much acute as it is existential. BloombergNEF expects that primary copper production can increase about 16% by 2040. That increase, needless to say, is rather short of demand. By the early 2030s, copper demand could outstrip supply by more than 6 million tons per year.[10]

However, according to the U.S. Geological Survey in 2019, construction ate up 43 percent of copper.[11] According to Copper.org, the average home has 88.5 kilograms (195 pounds) of copper wiring, fifty electric outlets, and fifteen to twenty switches.[12] The wiring is designed to carry 1,800 watts of power to incandescent lightbulbs, vacuum cleaners, and TV sets.

Except the light bulbs now draw 10 watts. The TVs are thin LED screens. The vacuum cleaners are battery powered. Everything we plug in has a little wall-wart to convert the power from 120 volts AC to low-voltage DC. Most of that 195 pounds of copper are superfluous, as are all those little wall-warts because you can now wire your house with computer cables and just plug in a USB cable. As the building codes catch up, we will soon no longer have 120-volt wiring that can cause fires, electrocute children, or waste 195 pounds of copper when a tenth of that will do. Our home wiring is being ephemeralized, and with it, a major source of demand for copper.

Copper isn't the only material that we are running out of. Geologist Simon Michaux, working for the Geological Survey of Finland (GTK) did an assessment of the minerals and materials needed to replace fossil fuels and concluded that the quantities are enormous, and the timing is almost impossible.

> Current expectations are that global industrial businesses will replace a complex industrial energy ecosystem that took more than a century to build. The current system was built with the support of the highest calorifically dense source of energy the world has ever known (oil), in cheap abundant quantities, with easily available credit, and seemingly unlimited mineral resources. The replacement needs to be done at a time when there is comparatively very expensive energy, a fragile finance system saturated in debt, not enough minerals, and an unprecedented world population, embedded in a deteriorating natural environment. Most challenging of all, this has to be done within a few decades.[13]

Michaux calculates that car batteries on their own would eat up half the world's lithium and nickel reserves. Furthermore, it gets harder and harder to get the metals out of the ore; there is less copper in a tonne of rock dug out of a copper mine today than there was left in the overburden of mines from a hundred years ago. This makes mining more expensive and more reliant on energy, primarily diesel fuel. Many mining processes use vast amounts of water; in his report "The Mining of Minerals and the Limits to Growth," Michaux notes that "there is an inverse exponential relationship with water consumption and ore grade."

We also don't have enough electricity to run everything. If we just try to electrify everything, we would need an additional 37,670 terawatt-hours, which "translates into an extra 221,594 new power plants."

Michaux concludes that this just isn't going to work. "The existing renewable energy sectors and the EV technology systems are merely stepping stones to something else, rather than the final solution. It is recommended that some thought be given to this and what that something else might be." In discussion on the GTK, they note that "the logistical challenges to replace fossil fuels are enormous. It may be so much simpler to reduce demand for energy and raw materials in general."

But long before there was Buckminster Fuller or Simon Michaux, there was William Stanley Jevons, who wrote in *The Coal Question*:

> There is hardly a single use of fuel in which a little care, ingenuity, or expenditure of capital may not make a considerable saving. But no one must suppose that coal thus saved is spared—it is only saved from one use to be employed in others, and the profits gained soon lead to extended employment in many new forms.

So, if we make things more efficient, we get more profits and more consumption instead of less. Our houses and cars all got bigger, and we used our savings to get more stuff with which to fill them.

But there are two sides to every equation. Jevons also noted that when a resource gets more expensive, we use it more carefully. "It is true that where fuel is cheap it is wasted, and where it is dear it is economized." Thus, when oil prices went through the roof in the seventies, we got the pivot to Pintos and small, fuel-efficient cars, and the Japanese invasion. When oil prices went down and engines got more efficient, we got SUVs.

Jevons works both ways. As lithium supplies tighten and the costs go up, batteries get new chemistries that use less lithium and get greater energy density. They are also being used, as Jevons predicted, in many new forms: we have seen the explosion in micromobility, with e-bikes, scooters, and minicars doing the same job of moving people with a fraction of the materials.

This is ephemeralization in action. We started this book noting that an iPhone has life cycle carbon emissions of 80 kilograms (176.4 pounds), but imagine the amount of materials and carbon emissions from making the stereos and computers and phones and VCRs it replaced.

Ephemeralization will happen naturally as the prices of raw materials increase. In the grocery store, it's called "shrink-flation," where the portions sold get smaller as the prices of ingredients increase. We will likely see this in transportation, with the inevitable changes driven by the economics of copper and lithium. After years of dramatic decline in the price of batteries, they were up 7 percent in 2022.[14] At some point, car makers will realize that concentrating on giant electric pickup trucks was a mistake. At some point, cars might ephemeralize from four wheels to two for most people as electric bikes continue to boom. Builders and architects will come to realize that Bucky Fuller was right, that how much your building weighs matters.

Ephemeralization is a technological form of sufficiency, redesigning and reinventing technology so that we can use less to perform the same function, whether it is powering our lights or getting from A to B.

Frugality

Benjamin Franklin defined frugality as a virtue: "Make no expense but to do good to others or yourself; that is, waste nothing." A dictionary definition is "the tendency to acquire goods and services in a restrained manner, and resourceful use of already owned economic goods and services." Frugality often comes up in discussions of personal finance, with recommendations that one skip the latte or the avocado toast. The recommendations often come from people who already have money and homes. In an article titled "Frugality Isn't What It Used to Be," Joe Pinsker noted, "Living a pared-down lifestyle necessarily means having a lifestyle to pare down. More often than not, the decision to live frugally is one made by people who can afford to opt out of a well-paid, well-spent lifestyle they have already secured."[15]

But there is another use of the term frugality that is more relevant to the question of the design and manufacture of stuff. It is called frugal design, frugal innovation, or frugal engineering, a term invented by Carlos Ghosn of Nissan Renault before he was arrested in Japan and escaped, carried out in a suitcase, who noted in 2006: "Frugal engineering is achieving more with fewer resources." Ghosn learned about it in India, where it has been practiced for decades in the design of everything.

It's not just about being cheap, although cost is a big part of it; there are other attributes that make frugal design attractive to anyone.

Robustness. India has a harsh environment with extremes of temperature, erratic electricity supply, and dust.

Portability. It's a big country with lousy transportation links, so you want small products that are easy to carry.

Defeaturing. This is one of the most interesting attributes. According to an article in the *Ivy Business Journal*, "Defeaturing consists of feature rationalization, or 'ditching the junk DNA' that tends to accumulate in products over time. With Indian consumers, firms can avoid implementing features that do little to enhance the actual product."[16]

Service Ecosystems. "Today, it is easy to see a plethora of small repair shops and other businesses that have mushroomed around population centres in India. The use of these service ecosystems—which comprise not just parts and repair but financing as well—can help firms enlarge their product markets."[17] The products are designed for serviceability because that is what the customer expects. In North America, the customer may expect it, but good luck getting it; I have spent considerably more on getting service on our clothes washer than I paid for it in the first place.

Ghosn applied the principles of frugal engineering and the mantra "do more with less" to produce low-cost vehicles that became hugely popular in Europe and Asia. And while Ghosn's career ended in ignominy, frugal engineering and innovation remain popular, with organizations such as Finland's Innofrugal promoting "the creation of sustainable economic growth via co-creating good quality, accessible, affordable, and sustainable solutions."[18]

Frugal design—doing more with less—makes sense in these times. Frugal engineer Balkrishna C. Rao says, "Frugal engineering is an important tool for tackling the challenges thrown by climate change and other planetary and manmade crises of our time. Frugal engineering is significant for all-round sustainable development."[19]

Frugal engineering started with cars, but it can be applied to anything. The other product that all the articles from a decade ago talk about is the ChotuKool refrigerator, a small fridge that cooled with a chip instead of a compressor. It and the Tata Nano, the first car designed from the ground up around frugal engineering principles, flopped in the market. According to Knowledge at Wharton, "aspirations matter."[20]

> Analysis showed that the "world's cheapest car" wasn't really an attractive positioning for what would have been an aspirational purchase for first-time car owners. The fridge failed because, as one of the senior executives responsible for ChotuKool noted: "We realized that the aspirations of the lower income people come from the richer people, and unless the rich buy, the lower income segment won't."

A decade ago, I wrote about both products, thinking the Nano was the right size for urban driving, and that "small fridges make good cities." This reinforces Joe Pinsker's point about frugality being a choice made by people like me who already have money and can walk to the neighborhood shop for milk and don't need to drive to Walmart. Others need or aspire to bigger cars and bigger fridges.

Aspirations do matter, which is one of the problems in dealing with the climate crisis, but it doesn't negate the value of frugal engineering. In many ways, it is close to what has been called simplicity.

Simplicity

Occam's razor is a thought exercise attributed to William of Occam, a theologian who died in 1327, and is usually interpreted as "the simplest explanation is usually the right one." It's called the razor because it "shaves away" complications. However, what might be the original source of the phrase, from his *Summa Logicae* in 1323, is perhaps even more relevant; *Frustra fit per plura quod potest fieri per pauciora*, or, "It is futile to do with more things that which can be done with fewer." Or, as Mies van der Rohe put it, "less is more."

I learned the phrase "radical simplicity" from engineer Nick Grant, who described how simple forms and appropriately sized windows make efficient Passivhaus buildings more affordable. He wrote:

> Passivhaus advocates are keen to point out that Passivhaus doesn't need to be a box, but if we are serious about delivering Passivhaus for all, we need to think inside the box and stop apologizing for houses that look like houses.

Simplicity is hard. It is much easier to add stuff and bling than it is to apply the razor and make it as simple as possible. This is true of everything, even writing; the French philosopher Blaise Pascal famously apologized: "I would have written a shorter letter, but I did not have the time." It's true of industrial design—Dieter Rams pulled it off for Braun in the sixties and seventies, noting in the last of his ten design principles:

> Good design is as little design as possible. Less, but better—because it concentrates on the essential aspects, and the products are not burdened with non-essentials. Back to purity, back to simplicity.

Do you remember the shock of first seeing an iPhone? Steve Jobs was a fan of Dieter Rams. Blackberry addicts like me agreed with TechCrunch, which published a post titled "We

Predict the iPhone Will Bomb": That virtual keyboard will be about as useful for tapping out emails and text messages as a rotary phone.[21] Meanwhile, Apple developed a design aesthetic of simplicity and minimalism that everyone now copies.

Architect Michael Eliason notes the many virtues of buildings that are simply designed. He calls them "dumb boxes."

> "Dumb boxes" are the least expensive, the least carbon intensive, the most resilient, and have some of the lowest operational costs compared to a more varied and intensive massing.... Every time a building has to turn a corner, costs are added. New details are required, more flashing, more materials, more complicated roofing. Each move has a corresponding cost associated with it.... Dumb boxes are great from an energy consumption standpoint because they're more efficient owing to lower surface area to volume ratio over buildings with more intensive floor plans.[22]

But designers in North America are terrible at it. In buildings, you get "sample-board modern" with twelve different materials applied to jogs and bumps and way too much glazing. They go green by adding expensive batteries and solar panels instead of insulation. In cars and trucks, you get front grills that look like giant chrome cheese-graters and six video screens inside. They go green by loading in batteries that weigh as much as a small car.

For all of the magazines and books about minimalism, people don't appear to want simplicity.

But as Andy Simmonds and Lenny Antonelli wrote, it's key to reducing impact:

> Designing and building as simply as possible—true value engineering or "integrated design." Asking: can this building be made simpler and more modest? Can the building be smaller or use less materials through innovative engineering approaches?"[23]

Flexibility

The first Volkswagen Type 2 bus rolled off the assembly line in March 1950. It was a simple box with an air-cooled engine at the rear, the driver way up at the front, and a lot of room in between. It came in different versions: a passenger model, a panel van, and a pickup truck by plopping different boxes on top of the standard base. It was a flexible design, cheap, and easy to maintain.

However, what happened to these buses after that was a cultural phenomenon. People installed sinks and beds, turning them into campers. They could carry anything, including lots of hippies to rock concerts and rallies. Roger White, curator of transportation at the Smithsonian, said, "It became popular with people who were rejecting mainstream American culture. It was their way of saying, 'We don't need your big V8 cars.'"[24]

It was such a successful design that VW kept making the original model until 1967 and in modified versions until 1999, when the engine was moved to the front for crashworthiness.

What made this such a success? It was designed for flexibility. It was adaptable to many different uses. It was simple; as the 1962 ad said, "You want to move something? Get a box." It wasn't very fast and had trouble on hills, but it was enough.

But there can be a cost to flexibility. The famous architect William McDonough's firm recently completed an office building designed to be flexible, with long spans between beams so that it could be converted to residential uses in the future if required. However, to get those long spans, the wooden beams got huge. They were already paying a premium in fiber to build out of mass timber, and now their search for flexibility chewed up even more fiber. Flexibility can be a double-edged sword.

Historically, everything was flexible and multifunctional. Witold Rybczynski explains in his book *Home* that "people didn't live in homes so much as camp in them" in the Middle Ages. Sigfried Giedion wrote in *Mechanization Takes Command*, that these were times of "profound insecurity, both social and economic, constraining merchants and feudal lords to take their possessions with them whenever they could, for no one knew what havoc might be loosed once the gates were closed behind him. The deeply rooted in the French word for furniture, *meuble*, is the idea of the movable, the transportable." If people dined at a table, it was made of boards on trestles—hence the boardroom—and taken down after a meal—hence turning the table. There were few chairs reserved for special people—hence the chairman. Everyone else sat on cushions placed on the trunks where belongings were stored.

Rooms were also flexible and multifunctional. Robert Kronenburg writes:

> Once we domesticated animals, we still moved according to seasonal grazing, and when humans finally settled to longer-term habitation, forming villages, towns and cities, the few rooms each dwelling possessed were multi-functional—used for sleeping, eating, entertaining and sometimes work. Consequently, they were furnished with demountable tables (that also served as beds), stools and benches (that also served as beds), chests containing clothes that also served as seats (and beds!). In Europe, it is only in the last three centuries that rooms with dedicated functions and associated specially designed furniture have appeared.[25]

Even when we got rooms, they were not very private; there were no halls, and all the rooms were lined up enfilade with one room leading into another. Judith Flanders writes that "for most of human history, houses have not been private spaces, nor have they had, within them, more private spaces belonging to specific residents, nor spaces used by all the residents in turn for entirely private functions." Trades worked from home or lived where they worked; there was no separation.

This all changed in the West in the eighteenth century, especially if you had money. With the Industrial Revolution, we got the factory and the office and the separation of work from home. We got parlors at the front of the house and bedrooms upstairs. Kitchens were separate from dining rooms, which were separated from living rooms. In my own house, a developer home from 1918 in a streetcar suburb, the dining room had sliding doors so that it could be separated from the living room while the servant prepared the table from the separate kitchen. There was even a steep stairway from the kitchen to a stair landing midway to the second floor so the servant wouldn't be seen in the front hall or living room if guests were visiting. There was one toilet for everyone; previous owners had ripped out that servant stair and put a toilet there on the ground floor.

But the main stair of the house was on the side, and the rooms on the second floor mirrored those on the first. So, when the kids moved out, we could divide the house into two separate units. The bedrooms on the second floor became a combo open living, dining, and kitchen area. My daughter's family now lives there, and my wife helps care for our grandchildren. It is likely that in the future, they will be taking care of us. This is the wonder of flexible design—a living space can be anything.

Robert Kronenburg notes that the way we live has changed:

> A flexible approach to our domestic environment is now
> necessary for a wide range of reasons; twenty-four-hour
> work patterns based in the home; changing family size
> and groupings; ecological issues that are questioning the
> desirability of commuting; lifestyle issues that envisage
> a more fulfilling personal life; the possibility of remote
> working due to communications technology.

Many nineteenth-century warehouses were flexible and have been converted into lofts and offices. Now, many are trying to convert older office buildings to residential since the pandemic has made them superfluous. It's not proving to be easy in many cases, mostly because they were lousy offices with floor plates that were too big. In Europe, where there were strict rules on how far workers could be from windows, conversions are far easier. It turns out that people's needs and desires are the same in either offices or homes: fresh air and a view of the outside. But in the North American office, those needs could be ignored. So instead, we got inflexible buildings with monster floor plates that will be difficult and expensive to convert.

Circularity

My dad's first job was in the circular economy. He married into a family that had what they called an "auto parts" business but could be more accurately described as the city's biggest scrapyard. If they got a call that a car was ready to be scrapped, he would dash out as fast as he could to get the battery; it was full of lead and the most valuable, portable part of the car. The tow truck would come by later to get the rest.

Labor was cheap relative to the cost of the parts, so it paid to disassemble the vehicle and sell the parts. It was the same with clothing and rags: the rag picker would visit your house and pick them up, and what couldn't be fixed and resold was turned into fine paper. High-quality drafting paper still is.

Everything used to be circular. In Japan and China, poop would be put in terra cotta jars and left out for pickup for use as fertilizer. It had value; rich people got paid more because they ate better food and produced better poop.

Even our pee was circular. In Ancient Rome, it was used in laundries and to whiten teeth.[26] Later, it had even greater value. The urea in urine breaks down into ammonia, which reacts with oxygen in the air to make nitrates. Mix that old pee with ash, and you get potassium nitrate, or saltpetre, a key ingredient of gunpowder.

Our milk, our Coke, and our beer were circular. They all came in glass bottles that were returned, washed, and refilled. There were no litter bins because there was no litter; kids would pick up bottles and take wagons full of them to the store for the deposit. You drank your coffee from a cup at a diner and left it there to be washed and reused.

Then came what I have called the "convenience industrial complex" based on the car, which became a kind of mobile dining room, and the development of plastics, made from fossil fuels, which were much cheaper than the natural materials they replaced, so cheap that they were disposable. We got the shipping container and globalism, which drove down the costs of production so much that it no longer made much sense to fix things. As Emrys Westacott wrote in *The Wisdom of Frugality*:

> There was a time when it almost always made economic sense to repair an item rather than replace it, so people would darn socks, patch sheets, and take their defective video recorder in for repair. But when half a dozen socks cost what a minimum-wage worker can earn in less than an hour, and when the cost of repairing a machine may easily be more than the price of a new one, some of the old ways can seem outdated.

So, we now live in a world where we drive cars powered by fossil fuels to buy food wrapped in single-use plastics, then to the mall to buy polyester socks that are too cheap to repair and fast-fashion clothing that for many people is too cheap even to wash, all shipped over from Asia in ships powered by dirty residual fossil fuels that are alone responsible for 3 percent of global emissions.[27]

In 1925, Calvin Coolidge said, "The business of America is business." But today, the business of America is turning fossil fuels into stuff and designing everything so that we need more stuff all the time, and then ensuring that millions of fossil-fuel powered vehicles have to be driven around by shoppers and companies promising two-hour delivery. It is as if the entire system was designed to maximize the consumption of fossil fuels. To paraphrase Robert Ayres: The economic system is essentially a system for extracting, processing, and transforming fossil fuels into energy embodied in products and services. And in the process, most of that embodied energy turns into upfront and operating carbon emissions.

Enter the new circular economy, pitched by the Ellen MacArthur Foundation as "a systems solution framework that tackles global challenges like climate change, biodiversity loss, waste, and pollution." They note, rightly, that it all starts with design.

> By shifting our mindset, we can treat waste as a design flaw. In a circular economy, a specification for any design is that the materials re-enter the economy at the end of their use. By doing this, we take the linear take-make-waste system and make it circular. Many products could be circulated by being maintained, shared, reused, repaired, refurbished, remanufactured, and, as a last resort, recycled. Food and other biological materials that are safe to return to nature can regenerate the land, fuelling the production of new food and materials. With a focus on design, we can eliminate the concept of waste.

None of this is new—our grandparents did it—but it has become the new buzzword; everything is going circular. The plastics industry in particular sees it as its savior; now that everyone knows that recycling is broken, they have hijacked the concept of the circular economy with what they call "chemical recycling" where they break plastics down to their basic chemical constituents and make new plastics that are indistinguishable from the originals. A report, "Accelerating Circular Supply Chains for Plastic" from Closed Loop Partners claims:

> There are at least 60 technology providers developing innovative solutions to purify, decompose, or convert waste plastics into renewed raw materials. With these available technologies, there is a clear opportunity to build new infrastructure to transform markets. These solutions can also help to decrease the world's reliance on fossil fuel extraction, lower landfill disposal costs for municipalities, and reduce marine pollution.

The Ellen MacArthur Foundation's approach is more sophisticated, calling for the reduction in the use of single-use plastics and demanding that "No plastic should end up in the environment. Landfill, incineration, and waste-to-energy are not part of the circular economy target state," and, "Businesses producing and/or selling packaging have a responsibility beyond the design and use of their packaging, which includes contributing towards it being collected and reused, recycled, or composted in practice." But all this collecting and reuse by somebody else and recycling takes energy and releases carbon. The sixty technologies pitched by the circular plastic people take vast amounts of energy to break plastics down to their molecular components; it's probably more than just making the virgin stuff.

In most circular economies, there is a form of entropy happening with the laws of thermodynamics in play. The aluminum people will keep saying that it can be recycled infinitely, but impurities in the aluminum mean that it can't be used for everything, and it still takes energy to do it. Paper deteriorates as the fibers get shorter. PET gets contaminated, so it cannot be turned back into water bottles. The more complicated the chemical recycling process gets, the more fuel it needs and the more carbon it releases.

You can't fight the laws of thermodynamics; it took energy to put the original product together, and because of the tendency to increasing entropy and disorder, it takes more energy to make it circular. As noted by Jouni Korhonen:

> Because of entropy, like all material and energy using processes, circular economy promoted recycling, reuse, remanufacturing and refurbishment processes too will ultimately lead to unsustainable levels of resource depletion, pollution and waste generation if the growth of the physical scale of the total economic system is not checked.[28]

The circular economy cannot exist in a growing economy; it just takes too much to run. It wants to stay linear because that is how the universe works: things break down to lower energy, disorder, and waste. The only thing we can do is slow down the process; as Korhonen notes, "the second law of thermodynamics means that every circular economy-type process or project should be carefully analyzed for its (global) net environmental sustainability contribution. A cyclic flow does not secure a sustainable outcome." We must make choices, use less, design for repairability and reuse, and stop pretending that recycling is circular; the recycling industry has just co-opted the term.

Universality

During a recent visit to the Gazelle bicycle factory and showroom in Dieren, Netherlands, I was bemused by the vast number of models that seemed to have relatively minor and subtle differences. After riding a lovely bike for the day, I accosted one of the product managers to complain that they should design a bike for boomers—lower, easy to step into, perhaps with smaller wheels. He acknowledged my complaints and took me over to see their "Easyflow" model, designed for people with disabilities. I took one look at it and decided that it was the bike of my dreams—the universal e-bike.

Universal design is a concept described by Ron Mace, who coined the term as "not a new science, a style, or unique in any way. It requires only an awareness of need and market and a commonsense approach to making everything we design and produce usable by everyone to the greatest extent possible." It was popularized by retired housewares manufacturer Sam Farber, whose wife had arthritis and had trouble using a standard potato peeler. He designed one with a big fat soft plastic handle that she could use. He didn't expect to sell many of them, given that it cost three times as much; it flew off the shelves because it was easier for everyone to use and was the foundation of the OXO line of easy-to-use products. Notably, the OXO company also says on their website: "Everything we make is built to last—our products are engineered for functionality and durability—it's why we guarantee them for life."

Key principles of universal design include:
- **equitable use:** the design is useful and marketable to people with diverse abilities
- **flexibility in use:** the design accommodates a wide range of individual preferences and abilities

- **simple and intuitive:** use of the design is easy to understand, regardless of the user's experience
- **tolerance for error:** the design minimizes hazards and the adverse consequences of accidental or unintended actions

Which brings us back to bikes and buildings. We know that one of the best ways to reduce life cycle emissions is to extend the life cycle. If I keep my iPhone for four years instead of two, I cut my emissions in half by avoiding getting another phone. Similarly, if we design our products and our homes so that they work for everyone, then everyone can use them for a longer time. I switched from a bike without a top tube to a "step-through," which used to be thought of as a "ladies' bike" because they are easier to get onto and off of. But I could see switching again to a lower bike with smaller wheels as I age out of my current bike.

The Gazelle people were shocked when one of our group, a young woman, said she wanted to ride it; she thought she would be more comfortable and confident. She could also ride that bike for life.

It is much the same with our homes. Aging baby boomers are spending millions to retrofit their homes for aging in place when it wouldn't be costing them anything had their homes been sensibly designed in the first place. Many spend all this money without realizing that one of the first things to go is the ability to drive safely, so they should think of how they will live without a car. They will not want a home so big that it is difficult to maintain, in a place where they cannot walk to get a loaf of bread. Transit expert Jarrett Walker has noted that "the unique feature of a city is that it doesn't work for anyone unless it works for everyone." The same should be said of our homes and our bikes.

Universal design reduces upfront carbon because it is like that Good Grips potato peeler: you can use it forever, and it is designed to last forever. You only buy one. It's not a great economic model, as marketing consultant Victor Lebow noted fifty years ago:

> Our enormously productive economy...demands that
> we make consumption our way of life, that we convert
> the buying and use of goods into rituals, that we seek our
> spiritual satisfactions, our ego satisfactions, in consump-
> tion.... We need things consumed, burned up, worn out,
> replaced, and discarded at an ever-increasing rate.

That worked just fine when we could keep burning fossil fuels to run that enormously productive economy. Today, we have to buy less and buy better, and with universal design, we will be able to use it much longer.

Resiliency

Every time there is a hurricane in the United States, we see photos of the aftermath where people are stripping their houses of the carpeting, insulation, and drywall in huge quantities. Steve Mouzon of the Original Green wrote:

> They call that boring white stuff we put on our walls
> "drywall" because so long as you keep it dry, you have
> a wall. But just as soon as it gets wet, it turns to messy
> mush. And even if it doesn't fall apart, it loves to host
> mold and mildew and make your family sick.... We need
> to learn how to build durable and resilient buildings like
> our great-grandparents did so that the summer shower is
> no reason to call the insurance adjustor; you simply wipe
> down the walls that got wet and never give it a second
> thought.

I have always hated drywall. If you join my wife in her little reading room/den (it was supposed to be her home office, but she got laid off before it was finished), you'll see that one wall is exposed brick from the former rear of the house and the ceiling is plywood. I am writing this from my home office, all exposed concrete block and plywood. In our cabin, there is one wall of drywall installed for acoustic reasons; it is covered in plywood.

But Mouzon, who lives in the southern United States, is not complaining about aesthetics. He is concerned about how it turns into moldy mush. Southern houses used to have wood-paneled walls, often cypress wood, which would just dry out. Alas, the most flood-resistant insulations are closed-cell foam. According to a study done after Hurricane Harvey:

> Open cell SPF, fiberglass batts, and rock wool all absorbed significant amounts of moisture. The fiberglass batts absorbed over three times the weight of the dry insulation (330% moisture content by weight) while the rock wool absorbed almost half the dry weight (42%). The samples of open cell SPF absorbed over 300 times the insulation dry weight (3040%) for one sample and 130 times (1320%) for the other. The fiberglass and rock wool had dried to 0% moisture by weight within eight days and four days respectively. After eight days, the open cell SPF samples retained about one third of the initial moisture (960% and 380%). These samples did not reach 0% moisture content for three weeks.

Spray foam has its own issues. It sticks to the studs and cannot be removed easily, so framing and sheathing cannot dry out. The study recommended instead that walls be designed to have removable interior components, such as wainscoting, so that they can be opened up and wet insulation removed.

The study also notes that Houston has had three "500-year floods" in the three years prior to the study. "Based on conversations with the owners and contractors we met, a large number of the affected homes have flooded several times." There is a lot of upfront carbon in all that drywall and insulation, not to mention a lot of money from insurance companies.

It costs more to build a resilient building, and it probably emits a bit more upfront carbon. But at least you only have to do it once.

Satiety or Enoughness

"If you have a garden and a library," wrote Cicero, "you have everything you need." This phrase is often repeated without acknowledging that Cicero was one of the richest men in Rome, had a huge garden managed by many slaves, and had a big bibliotheca in each of his three residences, the Roman one being one of the biggest houses in the city.[29] Cicero's definition of what he needed is probably different from yours or mine.

Defining what is enough and how much one needs is going to become a critical issue in the discussion of upfront carbon. In a discussion of a study on household energy use and well-being, Dr. Martiskainen told Carbon Brief:

> It's also really important that we start to have these discussions on what is the socially acceptable level of energy consumption.... What kind of society do we live in if we have people that have massive amounts of excess energy consumption [who] will then make climate change a lot worse for everyone else, versus people who can't afford to heat their homes?[30, 31]

It is not just a socially acceptable level, it is a mathematical reality; there is a carbon ceiling that we are trying not to crash through to stay under 1.5 degrees of warming. Some have suggested that we should each have a carbon ration. Vancouver writer Eleanor Boyle wrote in the *Toronto Globe and Mail*, that the climate crisis is like a world war, so let's start rationing.

> Rationing would change our lives and involve a word I've been trying to avoid: sacrifice. But what are we to do? Science shows we have barely 10 years to avoid disaster, suggesting we shouldn't count entirely on technological innovation or self-moderation. Meanwhile, we're all in a lifeboat with just enough space for each of us. Should we really be complaining about not getting first-class seats if doing so would bump others? That's what we're

doing when we consume too much of the stuff that fuels climate change.[32]

One problem with a carbon ration is that it would have to take upfront carbon into account, which complicates it tremendously. For example, if my carbon ration for 2024 is four tonnes, I might want to sell my Subaru and buy a Tesla with zero operating carbon emissions. But the upfront carbon emissions of the Tesla are about twelve tonnes. Is the carbon amortized over the life of the estimated life of the car, or do I have to buy carbon credits on some market of somebody else's surplus ration? Or does it make more sense to keep driving the Subaru? Or would I be happy with an e-bike?

Instead of determining what is a socially acceptable level or what is a rationed level, is it possible to know what level of consumption can provide satiety, a feeling of having had enough, to be happy? To be sufficient? In their 2014 book *How Much Is Enough?* Robert and Edward Skidelsky try to figure out what drives our desires for more.

> This book is an argument against insatiability, against that psychological disposition which prevents us, as individuals and as societies, from saying "enough is enough." It is directed at economic insatiability, the desire for more and more money. It is chiefly directed at the rich parts of the world, which may be reasonably thought to have enough wealth for a decent collective life.

In the poor parts of the world, as we will note in our discussion of inequality, people do not have enough. The authors, like our degrowthers in an earlier chapter, blame capitalism.

> Capitalism rests precisely on this endless expansion of wants. That is why, for all its success, it remains so unloved. It has given us wealth beyond measure, but has taken away the chief benefit of wealth: the consciousness of having enough.

The Skidelskys conclude that we have to stop thinking about money and start thinking about what we need for a good life, and that money is just a means to it. "To say that my purpose in life is to make more and more money is as insane as saying my purpose in eating is to get fatter and fatter." Their proposals for giving everyone a good life:

- full employment, which can be achieved by cutting hours worked
- unconditional basic income
- ban advertising, which is described as "the organized creation of dissatisfaction"
- a steeply progressive consumption tax, with a top bracket of 75 percent for people who spend the most

The authors conclude:

> Unless we take a collective decision to get off the consumption treadmill, we will never get to the point of saying "enough is enough." And if we don't do that, we will go on wondering what all that extra money was for.

Electricity

"Electrify everything!" is the cri de cœur as the world moves to electric mobility and heat pumps. Saul Griffith and the Rewiring America crowd say it will be easy! They write in their manifesto No Place Like Home, "No 'efficiency' measures such as insulation retrofits or smaller vehicles have been assumed here. Same-sized homes. Same-sized cars. Same levels of comfort. Just electric."[33]

One reason they say it will be easy to electrify everything is that "the electrified U.S. household uses substantially less energy than current homes." That's because fossil fuels are so inefficient. According to the famous Sankey drawing from the Lawrence Livermore National Laboratory, two-thirds of the energy from burning fossil fuels goes up the smokestack or out the tailpipe. The Electrify Everything gang says we will be only

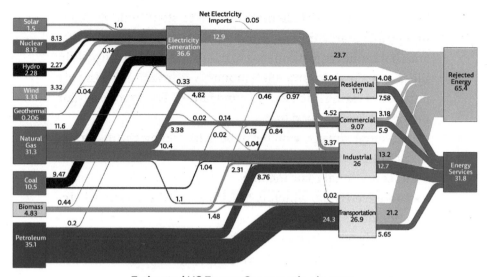

Estimated US Energy Consumption in 2021.
Credit: Lawrence Livermore National Laboratory.

using 42 percent as much energy when we go all-electric from renewables because none will be going up in smoke.

I have never understood this. We need exactly the same useful energy, or the exergy, not the amount of energy needed to make it. That was 31.8 quads (quadrillion British thermal units) running on coal and gasoline, and it's 31.8 quads running on wind and sunshine. There is no massive useful energy saving simply by going electric.

Energy expert Michael Barnard says much the same thing in an article titled "With Heat from Heat Pumps, US Energy Requirements Could Plummet By 50%."[34] Not really; not if you are measuring the output of a power plant, which is all that matters to the user. He explains: "What is the primary energy fallacy? Well, it's the belief that we have to replace all the primary energy inputs on the left-hand side of the chart above."

I believe that the energy fallacy is thinking that the energy inputs on the left-hand side are relevant for making anything other than CO_2 and waste heat. What matters, I reiterate, is the usable energy, the exergy.

So, let's briefly look at the math, using the Livermore chart as our base.

We need 31.8 quads. We are getting all of 7.11 quads from wind, water, and solar, and water is going down, not up. Nuclear isn't going anywhere fast, so we must find 16.56 quads of green electricity in a hurry.

Barnard does note that we will probably need fewer quads of exergy. He also suggests that heat pumps, which suck heat out of the air or ground, are more efficient than burning stuff to make heat, so when we convert from 3.94 quads of gas heating to electricity, it is not quad for quad, but about two-thirds less. So, if every building heated with gas is converted to heat pumps, we save about 2.6 quads, some of which will be eaten up by heat pump owners discovering that they now have air conditioning.

The point here is that there has been no miraculous discovery that we only need a third as much energy. The United States needs around double the low-carbon electricity it generates now, or three times as much renewable energy if nuclear is not expanded. Finding 13.96 quads of renewables is not nearly as daunting as replacing the additional 65.4 quads that went up in smoke, but it is still a challenge.

More quads of power will also be needed to deal with intermittency—those times when the wind doesn't blow or the sun doesn't shine, seasonality—it takes way more energy to heat in winter than it does to cool in summer, and there is a lot less sunlight, and peak loads—when everyone comes home for dinner and turns everything on at about the same time. Those seasonal and peak loads could be huge.

One study titled "Inefficient Building Electrification Will Require Massive Buildout of Renewable Energy and Seasonal Energy Storage" concluded that "all of our building electrification scenarios resulted in substantial increases in winter electrical demand, enough to switch the grid from summer to winter peaking. Meeting this peak with renewables would require a 28×increase in January wind generation, or a 303× increase in January solar, with excess generation in other months."[35] Others point out that this assumes no storage and no interconnections; a good grid could move power from where the wind is blowing and the sun is shining. The study also notes that better buildings could reduce this significantly.

Indeed, what is lost when looking at the Livermore chart is that the electrical system is designed for peak loads. You can store natural gas in the ground or in the pipes to keep our homes warm and spin up the peaker plants to supply electricity, but with solar and wind there is almost no storage capacity. Imagine a polar vortex event: the coefficient of performance of heat pumps drops in half in cold temperatures, so it is working flat out. Everybody's heat pump is sucking electricity as they try to suck heat out of the cold air. It's dinner time, and the induction range is eating kilowatts. This is the peak that the electrical system has to be designed to cope with.

The only way to cope with this is to flatten the peak. We could make our electric cars part of the system and use them to store power; we could interconnect the world with high voltage direct current cables and heat Maine in winter with solar power from Arizona or Morocco. Or we could super-insulate our homes and buildings and turn them into thermal batteries, with the utility controlling our heat pumps and dialing them back when loads must be reduced. This is where efficiency matters.

Then there is sufficiency. There is the not inconsiderable matter of the upfront carbon of making all of the electric cars, heat pumps, solar panels, wind turbines, and batteries. Those are included in the industrial emissions that eat up a quarter of the quads and are difficult to decarbonize.

This is where I disagree with the Electrify Everything gang that says, "Same-sized homes. Same-sized cars. Same levels of comfort. Just electric." I keep remembering our mantra: Use less stuff.

In 1954, the chair of the US Atomic Energy Commission said, "Our children will enjoy in their homes electrical energy too cheap to meter." It has been taken out of context ever since, and many have noted that producing electricity is a small part of its total cost, with distribution being much more significant. Even back in 1955, others were saying that the cost of building reactors would outweigh the savings in fuel.[36] The first director of Canada's Chalk River research facility said, "We do not expect to produce a cheaper source of power than that derived from coal—it is likely, in fact, to be somewhat more expensive. What we are aiming at is to increase the total power available."

Renewables are different. Build a wind or solar farm, and there is a one-time burp of upfront carbon emissions, and then it produces almost zero-cost electricity for years. A study by Deepa Venkateswaran at Bernstein Research concluded that the steel tower was 30 percent of the upfront carbon, the concrete foundation 17 percent, and the blades 12 percent. The full life-cycle assessment determined that the wind turbine averages 11 grams (0.4 ounces) of CO_2 per kWh; natural gas comes in at 450 kilograms (992 pounds), and coal, 1,000 kilograms (2,204.6 pounds).[37] Other studies put it as low as 6 grams (0.2 ounces) per kWh, and manufacturers also note that they are starting to use greener steel and recyclable blades, so the number is just going to get lower. Similarly, solar panels get thinner and cheaper and more efficient.

But even with renewables, electricity will never be too cheap to meter. We still have to build out a distribution system that runs around the world; we have to overbuild the system and build storage to deal with intermittency and winter peaks. And we still have to use less electricity so that we can spread it around to the half of the world's population that still suffers from energy poverty, instead of hogging it all in the rich north. "Same-sized homes. Same-sized cars. Same levels of comfort. Just electric" is a fantasy. We still need sufficiency, and we still need efficiency. And it wouldn't hurt to also design for intermittency.

Intermittency

As noted earlier, our biggest problem is peak demand, and the best solution is to shave the peak. Efficiency will reduce the overall demand, although we should start with sufficiency and frugality. But a related approach is to design for intermittency. We know that there will be times when the wind doesn't blow, and the sun doesn't shine. Many are suggesting that this is a role for hydrogen; we make piles of it and store it when the sun shines and make electricity from it when it doesn't. But I keep falling back on the words of Dr. Steve Fawkes in his twelve laws of energy efficiency:

> An exciting energy or energy efficiency discovery in a lab somewhere is not the same as a viable technology, which is not the same as a commercial product, which is not the same as a successful product that has meaningful impact in the world.[38]

Perhaps the best approach is the old-fashioned one, how we have done it in the past. Kris De Decker of Low-tech Magazine reminds us that our ancestors had to deal with the intermittency of wind power on a regular basis:

> Because of their limited technological options for dealing with the variability of renewable energy sources, our ancestors mainly resorted to a strategy that we have largely forgotten about: they adapted their energy demand to the variable energy supply. In other words, they accepted that renewable energy was not always available and acted accordingly. For example, windmills and sailboats were simply not operated when there was no wind.

But sailors got good at figuring out global wind patterns, the combination of trade winds and westerlies that were relatively dependable. Similarly, we know when we are going to get peak demands, and we can prepare for it. After I first wrote about intermittency, Dr. Es Tresidder replied from Scotland that we would need lots of winter storage. "For example, at the moment we're in the middle of a long, very cold, low-wind weather period in the UK. In a future with lots of EVs and lots of heat pumps electricity demand will be high even with better buildings, demand response, and behavior change. So let's do all of those things, but also push for H_2." Hydrogen expert Michael Liebreich concurred: "We certainly do. Otherwise it's a case of the nation's heating working 99.8% of the time but the 0.2% of the time it doesn't, thousands of pensioners could die."

Liebreich is again talking about our peak demand, and proposing a vast network of caverns filled with green hydrogen as a backup for what he calls a 0.2 percent of the time event. This all seems rather expensive. Others have suggested that the cost of renewables have dropped so quickly that the answer is simply to overbuild. Mark Perez proposed in a study that intermittency could be overcome by oversizing the system by a factor of three.[39] This is also what is known as the Law of Large Numbers:

> The Law of Large Numbers is a probability theorem, which states that the aggregate result of a large number of uncertain processes becomes more predictable as the total number of processes increases. Applied to renewable energy, the Law of Large Numbers dictates that the combined output of every wind turbine and solar panel connected to the grid is far less volatile than the output of an individual generator.[40]

But I keep weighing all these solutions in terms of upfront carbon, and they involve megatonnes of emissions from the construction of an entire hydrogen infrastructure, or three times the amount of renewable energy plus the mining of untold tonnes of resources. Learning from our ancestors, perhaps a better approach is to try to reduce our demand by a factor of three.

When it got really cold, our ancestors put on a sweater and dressed appropriately; clothing was the best method of dealing with intermittency since our bodies are so effective at turning food into heat. Today, we can put a sweater on our houses with good insulation. When they started building homes around fireplaces and then central heating, they built them small and square to minimize the area that needed to be warm; moving firewood and shoveling coal was time-consuming and expensive. We have become accustomed to big homes and cheap dependable energy; if we are to design for intermittency, we would also make them small, square, and super-insulated, and would share walls with our neighbors.

Our ancestors knew that if they wanted to sail dependably, they had to go to where the wind was; we have to do the same and connect to where the wind and sun are. Es Tresidder's fellow British citizens all fly south to Spain and Morocco; that is where the sun is. A good HVDC wire could connect Scotland to Morocco and bring that sun to him.

A look at the Energy Information Agency chart of how energy is used in the United States illustrates why we can deal with intermittency without having to resort to caverns full of hydrogen. Industrial users are already used to dialing back production when energy input costs are high; they do it to save money now as a matter of course. The biggest section of residential demand, "all other uses," includes washers, dryers, computers, and stoves, most of which do not need to run at peak demand times. Many people with time-variable electrical rates run them when demand is low. Water heaters are now being connected via the Internet to the utility meter to buffer peak demand. A recent Australian study concluded that "a heater with a 300-litre tank can store about as much energy as a second-generation Tesla Powerwall—at a fraction of the cost."[41] Well-insulated houses can also act as thermal batteries.

Cooking is an interesting issue. When Saul Griffith and others proposed selling induction ranges with giant batteries, I scoffed. They were proposing them because many houses do not have electrical supplies big enough to supply a 40-amp circuit to a stove. With the batteries, a stove could be plugged

U.S. electricity retail sales to major end-use sectors and electricity direct use by all sectors, 1950–2019

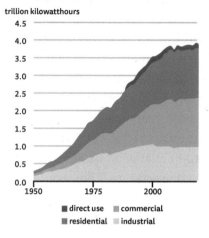

trillion kilowatthours

direct use ■ commercial
residential ■ industrial

Source: U.S. Energy Information Administration,
Monthly Energy Review, Table 7.6, July 2020

U.S. residential sector electricity consumption by major end uses, 2019

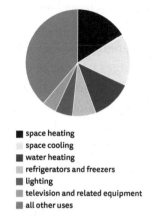

■ space heating
space cooling
■ water heating
refrigerators and freezers
■ lighting
television and related equipment
■ all other uses

Note: Space heating includes consumption for heat and operating furnace fans and boiler pumps. All other uses includes miscellaneous appliances, clothes washers and dryers, computers and related equipment, stoves, dishwashers, heating elements, and motors not included in other uses.
Source: U.S. Energy Information Administration, Annual Energy Outlook, 2020, Table 4, January 2020

US Electricity Sales and Comsumption by Major End Users
Credit: Energy Information Agency

into a standard 15-amp, 120-volt plug and charge all day, storing enough power to run the stove when dinnertime comes. But the battery-powered stove takes over when demand is at its absolute peak—dinnertime, when everyone is home and running everything, and the sun has gone down. I am reconsidering my position on this one in the light of the peak demand problem.

Refrigerators and freezers are a big slice of this electricity pie. Freezers could be part of the solution if people would return to chest-style freezers—the cold doesn't all dump out when you open the door, and a well-insulated freezer could keep food cold for days without electricity. A diet designed for intermittency would also help. The low-carbon seasonal diet, with lots of root vegetables and rice, doesn't need refrigeration.

There are some technologies that could be very helpful in dealing with intermittency; Sunamp makes a water heater based on change-of-state materials. Other companies are building them into heat pumps so that they store heat or coolth when energy prices are low. This is the oldest of technologies; our fridges used to be cooled with ice, the melting of which is the most common change of state that we use every day, sucking up 80 calories of heat energy for every gram that melts. Steam releases 540 calories of heat for every gram that condenses, which is why steam radiators were so effective. Modern change-of-state materials are mixed to change at useful temperatures, to keep a home at an even temperature. They would be excellent at shaving peak demands.

In short periods of peak demand, electric cars could be part of the solution, feeding electricity back into our homes and the grid. In longer periods of intermittency, they become part of the problem, with their huge demand. Here, once again, I call for sufficiency—we don't need such big electric vehicles, and we don't need so many of them. In a well-designed community of well-built homes, people can walk or bike.

What is the alternative? In Canada, we have lots of waterpower, and if we reduce demand, as I have proposed, we probably don't have to worry much about this in most of the country. In Europe, Michael Liebreich calls for hydrogen:

> ...stored in salt caverns, in pressure vessels, as a liquid in insulated tanks, or as ammonia. It will be moved around, cheaply via pipelines, or at a higher cost by ship, train, or truck. And it will need to be strategically positioned to cover the risk of supply shocks, whether they be the result of normal weather patterns, extreme weather events and natural disasters, conflict, terrorism or any other cause.[42]

Sheesh. The cost of this would be astronomical, especially since it is essentially a backup system. It's not realistic. Changing building codes so that our homes are super-insulated is. Fixing our cities so that we can walk or bike safely and get what we need within fifteen minutes instead of driving is. Having a cupboard filled with rice and a week's supply of food that doesn't need refrigeration is. Wiring up our heat pumps and water heaters so that they can be adjusted by the utility is. Putting on a sweater when the utility turns down the heat is.

Intermittency, or variability as some think is a better description, was part of our way of life for everyone until we got the steam engine and fossil fuels. People lived with it; they had no choice. It would not be such a terrible thing for us to get used to a bit of intermittency again, and to be prepared for it.

Operating Efficiency

My iPhone 11 Pro has an 11.91 Wh battery. Apple claims it will power the phone for up to sixty-five hours with audio only, and seventeen hours with video playing, and the whole thing weighs 188 grams (6.6 ounces). This is extraordinary efficiency because the customer wants a teensy lightweight phone that lasts a long time. In a phone, energy efficiency is palpable and directly affects the user.

With almost everything else, energy efficiency is not so highly valued. There are people still hoarding incandescent light bulbs because they prefer the quality of light, not caring that they use up to fifteen times as much energy. As cars got more efficient, people switched to SUVs and pickups. And even as building codes increased the energy efficiency of our homes, they got bigger.

But in a low-carbon, all-electric world, everything is an iPhone. Everything has to be as light and small and energy efficient as possible.

There are many who say that in an all-electric world with heat pumps and cheap renewable energy, that efficiency wouldn't matter; carbon emissions are our problem, and they don't have any emissions. I have said it myself with respect to building renovations; the cost and disruption required to make an older building as energy efficient as a new one is daunting. As engineer Toby Cambray eloquently put it, maybe a mix of insulation and heatpumpification is a better approach. Writing from the UK, he breaks my spellchecker with his verbing of heat pump:

> This does not however mean that it's a good idea to put a heat pump in a building with poor fabric efficiency. Although there are cases where other constraints mean we have little choice; ultimately, we need to both (mostly) Insulate Britain and (mostly) Heatpumpify Britain.

There are millions of existing houses and buildings that have to be upgraded, and the standard line we always hear is that they need to be demolished so that a more energy-efficient building can replace it. This argument loses all credibility now that we know about upfront carbon.

Some say the same logic applies to new buildings. Why spend so much money on insulation when our problem is carbon emissions, and there aren't any carbon emissions with heat pumps running on clean electricity?

And think of the upfront carbon in all those expensive European windows and the extra insulation! I keep saying that the Passivhaus standard should be the standard for every building, but as Skylar Swinford writes:

> Unfortunately, somewhere during the process of embodied carbon awareness-raising a misperception has taken hold: that if you're not careful, Passive House practice will do more climate harm than good; that the extra insulation and triple-pane windows on a Passive House can backfire.

But much depends on the size of the windows and the choice of the insulation. There doesn't have to be a backfire.

We hear the same argument with electric vehicles. They are so much better than gas-powered cars and trucks, but don't let the perfect be the enemy of the good! If people want a three-ton electric pickup that goes from zero to sixty in three seconds, let them have it! Efficiency doesn't matter!

But when winter heating switches from fossil fuels to heat pumps, the system has to be able to cope with the coldest, darkest days of the year. Coal is easy to store; you can pile it up anywhere. Gas is stored in caverns and huge amounts are in the pipes themselves, but also in the ground where it came from. This is what lets us live in marginally insulated glass boxes; as Reyner Banham wrote in 1969 in *Architecture of the Well-Tempered Environment*:

> For anyone who is prepared to foot the consequent bill
> for power consumed, it is now possible to live in almost
> any type, or form of house one likes to name in any
> region of the world that takes the fancy. Given this conve-
> nient climactic package, one may live under low ceilings
> in the humid tropics, behind thin walls in the arctic and
> under uninsulated roofs in the desert.

But imagine you are trying to live in this house in January. Imagine you want to also charge up your F-150 Lightning. We quoted that study in the last chapter that calculated that we need a "28× increase in January wind generation, or a 303× increase in January solar, with excess generation in other months;" that study also noted that if our buildings were highly efficient, we would only need "4.5× more generation from wind and 36× more from solar."

It is likely that it will not be so dire, and we will not need so much power; electric cars can provide a huge amount of storage if interconnected with our electrical systems. Water heaters are now being controlled by utilities to balance the grid; if our houses were all super-efficient then they could be used as thermal batteries, with the heat pumps turned off at the peak times. The IPCC notes that "in well-insulated buildings, switching off a heat pump for several hours can have little impact on indoor temperatures." In a Passive House design, it can take days for the temperature to drop significantly.

But if we are not all going to freeze in the dark in January, then we cannot forget about operating efficiency.

Operating efficiency matters with our cars and trucks as well. A vehicle is like an iPhone in the sense that how long it runs is extremely important to the owner; that's why they have the term "range anxiety." But because it's on wheels and not in our pocket, the manufacturers can just pile on more batteries.

So, an electric Ford Mustang has a giant 91 kWh battery pack that will push it 425 kilometers (264 miles) with an efficiency of 214 Wh/km. A Tesla Model 3 that seats the same number of people and takes them almost as far (380 kilometers or 236 miles) has a battery of only 57.5 kWh and an efficiency of 151 Wh/km.

But with public charging stations costing eleven cents per kWh, nobody cares about efficiency; it is still a quarter of the price of gasoline. In Canada, electric cars are rated in Le (litre equivalent)/100 km so that comparisons with gasoline cars are easier; our Subaru Impreza is rated at 8.5 litres/100 km. No wonder people are buying the biggest electric car they can afford; they cost nothing to run.

But we are talking operating energy here, not upfront carbon emissions, which are obviously going to be radically different. Again, it's one of those cases of when you look through the lens of upfront carbon, everything changes.

Increasing efficiency can also increase upfront emissions because everything gets bigger and heavier, thanks to the Jevons Paradox and rebound effect. That's why Samuel Alexander writes, "without sufficiency, efficiency is lost." Some, like Saul Griffith and the Electrify Everything gang, say if it is all-electric and powered by renewables, who cares about efficiency because it is carbon free? But efficiency is still necessary if we are going to minimize the amount of clean energy that we need.

Design Efficiency

Buckminster Fuller wrote, "How much does your building weigh? A question often used to challenge architects to consider how efficiently materials were used for the space enclosed." He aimed this at critics of his lightweight Dymaxion house of 1927. Fuller cared about this; Dymaxion is supposedly an abbreviation of "maximum gain of advantage from minimal energy input."

Architects never really cared about how much their buildings weighed; they had engineers design the footings and the structures, specifying how many tonnes of concrete and steel would be used. At least they never cared until the question of embodied carbon came up; now they care a lot. One reason they started working with wood was that it was lighter and would have smaller foundations, as well as being much lower in upfront carbon than other materials, but they didn't use it particularly economically; in fact, they thought using more of it was a good thing, storing more carbon. But as Lenny Antonelli and Andy Simmonds write, this is the wrong approach:

> Use natural resources extracted from our shared bio-sphere respectfully and efficiently to substitute for higher embodied carbon materials. Use as few materials as possible to achieve the design. Using a "renewable" material inefficiently, whether to "develop the market" or "store carbon" is wrongheaded—efficient use of the same quantity of material, substituting for higher carbon options across many projects, makes far more sense.

Fuller liked round buildings because they enclosed the most area for the amount of perimeter, but having owned a geodesic dome, you can trust me, they are hard to furnish and live in.

Chris Magwood and the Builders for Climate Action developed the BEAM tool to compare the upfront carbon emissions for different building materials and choices, but I used it for a magazine article to compare an "efficient" design to a typical house plan. I found that if I couldn't design it to be a circle or a sphere like Bucky did, I wanted it to be a cube, the simplest and squarest plan I could devise. I learned that design efficiency made a significant difference in the upfront carbon emissions—they were 30 percent less with the same useable floor area.

Design efficiency is a recognized term defined as "exhausting all techniques to optimize the design in its use of material quantities to provide maximum capacity at minimum cost." That sounds exhausting, and it sounds late, as if the building or product is designed already and then gets "optimized" or "value engineered" after the fact. In buildings, it's important to get those design efficiencies upfront. Structural engineer Mark Webster told Paula Melton of Building Green: "It would be great if architects would reach out earlier to us [structural engineers] to help them make decisions related to building form and structural materials." He added, "It's increasingly obvious, the role that we have to play in terms of embodied impacts with respect to climate change."[43]

Design efficiency, like frugality, often fails to meet consumer tastes. Bucky Fuller couldn't sell many Dymaxion houses, and Tata couldn't sell many Nano cars. If one designed an electric car efficiently today, it wouldn't look like a regular car; there is no need for that trunk in front where the engine used to be. It might look more like my beloved flexible Volkswagen bus. One company, Canoo, does this with a design that is very much a box on a platform. The designer said, "Cars always have been designed to convey a certain image and emotion; however, we chose to completely rethink car design and focus on what future users will actually need." Critics said, "Personally, I think it is as ugly as sin. Certain shapes are appealing to most people."

A problem that is inhibiting design efficiency is skeuomorphism. Tech journalist Clive Thompson explains: "A skeuomorph is a piece of design that's based on an old-fashioned object. You've invented a newfangled technology, but you design it to look and act much like the old tech it's replacing."[44]

The skeuomorph that drives me mad is the DSLR camera. The original single-lens reflex camera was designed around film rolling from one side of the camera to the other, with a mirror flipping up out of the way. It was a brilliant design for film, but it is awkward to hold and is an ergonomic nightmare.

There was an explosion of innovation when digital cameras were developed; they could be any shape. Nikon developed the wonderful Coolpix that twisted in the middle; it was easy to hold in front of you, up high, or looking down like you do on a Hasselblad. And nobody bought it because it wasn't skeuomorphic—it didn't look like a camera. Now, even when the viewfinders are digital, they all still look like film cameras with a flipping mirror. They are not efficient designs. As Clive Thompson notes, "skeuomorphs also wind up hobbling the new invention. Because skeuomorphs are based on the physical limits of an old-fashioned device, they get in the way of a designer taking full advantage of the new realm."

The key doctrine in the new realm is to use less stuff, and to design everything as efficiently as possible. That means eliminating the skeuomorphs. But I am also getting myself stuck in an old-fashioned age; for most people, the camera has disappeared into their phone, and for some, their electric vehicle is an e-bike. Now that is design efficiency.

Inequality, Inequity, and Justice

If one accepts that there are about five hundred gigatonnes of carbon budget left if we are to stay under 1.5°C, a big question is: who gets it? How do you divide it up? Or is this all going to be a global tragedy of the commons? According to OXFAM, the top 10 percent of global carbon emitters (including probably everyone reading this book) generate almost half of all greenhouse gas emissions. UN Secretary General Guterres says, "The climate crisis is a case study in moral and economic injustice."

The Climate Inequality Report 2023 calls for strict regulation on polluting purchases for the top 10 percent, hefty wealth and corporate taxes with cash transfers to the bottom 50 percent, and "carbon cards" to track and cap high personal carbon footprints.[45]

The study "Equity Assessment of Global Mitigation Pathways in the IPCC Sixth Assessment Report," the mitigation report that

I quote so often, says mitigation is all very well, but why are poor southern nations carrying the burden and not being allowed to improve their lives and get a bigger share of the emissions? They write that "The burden of climate change mitigation is placed squarely on less developed countries, while developed countries continue to increase their energy consumption unhindered by constraints on the use of fossil fuels."[46]

Authors Tejal Kanitkar, Akhil Mythri, and T. Jayaraman note that the IPCC bakes in larger emissions per capita in rich countries, including far greater consumption of fossil fuels.

> Our analysis of the regional trends underlying the global modelled scenarios in the IPCC's 6th Assessment Report indicates that not only do the scenarios not "make explicit assumptions about global equity," but they in fact project existing global inequities far into the future. The scenarios do not consider the differential energy needs of countries in the future based on their levels of development. Per capita GDP, consumption, and energy use remain significantly high in developed countries, even as most developing regions are projected to stay at very low levels of income, consumption of goods and services, and energy.

They note also that one should consider the historical emissions that got us here in the first place. By their calculations, using a 624 $GtCO_2$ budget, South Asia should get 153 Gt, because of their population and their entitlement to a higher standard of living, and North America should just get 31 Gt. Taking historical emissions into account, including the 509 Gt that the United States has poured into the atmosphere since 1850, South Asia should get 506 Gt and North America should have to go negative, sucking 351 Gt out of the air.[47]

The study was written by South Asians working at a climate change institute in Chennai, in a country where the carbon emissions are 1.9 tonnes per person, roughly a tenth of a Canadian, a country where the coal-fired electricity generation is expanding to run the increasing number of air conditioners that are becoming necessary for survival.[48]

According to the UN Emissions Gap Study, eliminating carbon inequality would require the top 10 percent to reduce their emissions by 91 percent, while allowing the poorest 50 percent a 300 percent increase. David Wallace-Wells notes in *The Uninhabitable Earth* that "there is something of a moral crime in how much you and I and everyone we know consume, given how little is available to consume for so many other people on the planet." But people would rather commit moral crimes than recognize the brutal present and historic unfairness of it all and give up their pickup truck or hamburger. We must, as Gandhi said with the utmost moral clarity, "live simply so others may simply live."

CHAPTER 3

STUFF

Carl Sagan said: "If you wish to make an apple pie from scratch, you must first create the universe." It is a statement about boundaries; does making it from scratch mean getting flour and apples from the store, or do you go back to planting the tree and the wheat, or to the development of photosynthesis, or to the miracle of carbon after the Big Bang?[1]

Similarly, Mike Berners-Lee had issues with boundaries, noting that "the ripples go on and on for ever." We have to follow those ripples to examine not only what a product is, but why it is. We know from Mike Berners-Lee that a single-use coffee cup has a carbon footprint of 110 grams (3.9 ounces). That's not much; the milk in a latte is considerably higher.

But I am more interested in why we have this single-use coffee cup, how it became part of our culture, and why it is so difficult to eliminate it. A look at a disposable coffee cup has to go way beyond the boundaries of a traditional life cycle assessment. When you expand the boundaries to what this cup actually wrought, from the drive-through coffee shop to the SUV with the twelve cup holders to the consumption of 20 ounces instead of 5, to the linear take-make-waste culture of convenience, you see that it is much more than 100 grams (3.5 ounces) of carbon embodied in paper and plastic.

That's why this is not a book about the life cycle assessment of everything, although I will admit that is how it started. The point of this section of the book is to look at the *why*, and to evaluate it all in terms of our strategies for sufficiency. We are looking at everything through the lens of upfront carbon, rather than calculating it.

The Single-Use Coffee Cup

When I attended the School of Architecture at the University of Toronto in the early seventies and wanted a cup of coffee, I would go across the street to Harry's, sit down at the counter, and drink five ounces of coffee out of a china cup. The coffee was small because real estate is expensive, and Harry didn't want me sitting there all day.

Disposable cups existed, but they were not very good; they would get soggy and leak if you didn't drink quickly. Cups made of polystyrene foam were introduced in 1960, and in 1964, a convenience chain based in Long Island, 7-Eleven, became the first to offer coffee to go in disposable cups.[2] At about the same time in New York City, where Greeks operated most coffee shops, the "anthora" cup, a misspelling of amphora, became legend.[3] The operators of these coffee shops that did a little takeout business on the side didn't realize how these cups would change their world and put many of them out of business.

Expanded polystyrene cups worked well but were an environmental nightmare, with pieces of foam littering cities and filling dumps. Also, in 1984, Solo invented the Traveler lid, the high-top sippy-cup lid we know and love today. But it only worked with paper cups. In 1987, while planning a major expansion of Starbucks, founder Howard Schultz decided to go with an upgraded paper cup with a polyethylene liner to prevent leakage and sogginess and a dome top that wouldn't mess up the foam on his Frappuccinos. He also presented this as a more environmentally friendly cup. If only.

Starbucks was originally presented as "the third place" after home and office, described as that "warm and welcoming environment where customers can gather and connect." But 74 percent of Americans live in the suburbs now, and they hang out in their vehicles. It didn't take long for Schultz and Starbucks to discover that takeout was more profitable. Because the customer supplied the real estate where they would sit and drink, the company could make more profit by selling bigger containers; the car makers struggled to keep up with their cup holder sizes. The pandemic finally killed the third place; now, 70 percent of the business is drive-through, and 90 percent of new stores will be drive-through-equipped.[4] That will generate a lot of cups, even more than the 50 billion single-use coffee cups used each year in the United States.[5]

It's not that these cups are not recyclable; they are, but it is different from normal cardboard recycling. The cups must be collected separately or picked out of the garbage, then shipped to a facility where they are shredded and soaked until the plastic separates from the pulp. The polyethylene has no value, and the paper is sub-food grade and isn't worth much either—certainly not worth all the trouble it takes to do it. A Toronto paper shredding company charges five cents a cup to customers who collect them from employees and bundle them with their other shredding. A recent CBC report noted that Tim Hortons cups are often shipped to India where they are pulped for the cardboard, which is turned into Kleenex; the plastic is landfilled if we are lucky.[6] Nobody is counting the carbon from the dirty fuel it takes to move a used paper cup eleven thousand kilometers (6,835 miles).

But the vast majority of cups are not collected; they go out the door with the customer and end up in the garbage somewhere, and then the landfill. It's a linear process that supports our culture of convenience. It's a profitable process; the takeaway is cheaper to run that sit-down service. Many companies have tried promoting refillable cups, better cups, recyclable cups, and tried to make the linear process circular.

With coffee cups, for example, there have been many attempts at putting RFID chips in them where people buy it at one store and drop it off at another, or even just giving people a reduction in price if they bring their own cup, which a very small percentage of the population will bother to do. But it is not enough to change the cup; we must change the culture and the entire concept of takeaway coffee.

The problem is that it is hard to bend a linear economy into a circle. It wants to stay linear. We have been trained through convenience and lack of alternatives to accept that the world is linear. Leyla Acaroglu has written that this is all designed to consume more:

> The systems of disposability permeating our lives are a product of economic incentives and the systems archetype of a race to the bottom, offering the cheapest price tag by the producer and the most convenient solution to the "consumer," at the cost of the society and the planet. But whilst it may seem like cheaper products are better for the consumer, the net gain is always skewed towards the producer.

Then there is the classic question of boundaries. It's not just the coffee cup that is the problem at 110 grams (3.9 ounces) of CO_2, it's the system. The average customer at the drive-through window waited 4.4 minutes in 2018; it is probably longer now. According to Natural Resources Canada, ten minutes of idling burns .25 litres (.07 gallons) of gasoline; 4.4 minutes burns .11 litres (.03 gallons). Burning a litre (.26 gallons) of gasoline releases 2.3 kilograms (5 pounds) of CO_2, so the average idle time waiting for that cup of coffee releases 253 grams (8.9 ounces) of CO_2, well over twice what making the cup purportedly releases.

The cup, the cup holder in the SUV that's big enough to be a mobile dining room, the need to always be sipping something, that's the system, the culture of convenience. That's why we have to change the culture, not the cup.

Writer Katherine Martinko has described drinking coffee in Italy:

> While travelling in Sardinia, Italy, my husband and I stopped at a small roadside bar for an early morning

coffee. The barista pulled our espressi with a deft hand and pushed two white ceramic cups and spoons across the counter, along with a little sugar dish. We stirred, drank it in a few gulps, and chatted briefly with the other people lining the bar, also enjoying a quick coffee. Then we headed back out to the car and continued on our way.[7]

This is a different coffee culture. I have been in Italian urban espresso bars that were no bigger than my desk—you sidle up, knock it back, and leave. Katherine notes it is healthier "as people transition from calorie-laden vats of coffee-like concoctions, full of sugary syrups and flavourings and mountainous whipped cream." It's cheaper, and there is no waste.

Insight: Sit down, enjoy the coffee, and refuse paper cups.

From the Two-by-Four to Mass Timber to a Block of Flats

The North American two-by-four is the core of the most ephemeral of building technologies; our typical home has been described as "an aggregation of toothpicks." Early settlers might clear their land, square up the wood, and build with what we now call timber frame, but as forests got cleared for farmland and lumber was shipped from farther away, it became necessary to standardize. By about 1900, lumber was usually two inches thick and boards, one inch. But the saw blades were not thin, and some wood was lost; the wood was also often run through planers and edgers, which reduced it further. In 1924, a standard was agreed to that allowed for different tools and technologies, with the so-called two-by-four really being a one-and-a-half by three-and-a-half. Vast western spruce and fir and southern pine forests made the standardized two-by-four cheap, and along with inexpensive nails, it made rapid expansion of housing across North America possible.

It is used in everything from starter houses to McMansions. Paul Andersen and Paul Preissner, designers of the US Pavilion at the 2021 Venice Architecture Biennale that was built out of two-by-fours told Kate Wagner:

> One of the things that we really like about framing is
> it's the same for everybody. It doesn't matter how rich
> or poor you are, your house—at least the structure of
> your house—is all made of the same stuff. No amount
> of money can buy you a better two-by-four than the one
> that's also in the crappiest house in town. I think that's
> really democratic, that everybody—you and Beyoncé—
> has access to the same materials, and they're the best
> materials.[8]

Wood frame construction is not without its problems or limitations. It has to be kept dry, or it becomes food for mold. Wood is not a terrible insulator, but it is not as good as the insulation stuffed around it, so it acts as a thermal bridge. When it is nice and dry, it burns easily. But its flimsiness and flammability have limited its usefulness, which is why building codes prohibited it from being over three stories high until recently. Wood structures could be protected with rated drywall assemblies and sprinklers, but the concrete, masonry, and steel industries lobbied hard to keep wood out of bigger buildings.

With the emergence of the climate crisis, it was quickly recognized that building with wood had a much lower carbon footprint than building with concrete or steel, although how much less is still open to question. But with lightweight wood framing, there was no question at all; there is just so much less weight. Foundations are smaller, so the structure weighs a fraction of concrete. Building codes in many jurisdictions were loosened to allow it in buildings up to six stories high, with sprinklers and full alarm systems. It was cheaper and quicker, too, allowing for greater economies. But it requires good detailing and care while building to control sound transmission and keep the water out. In much of the United States, residential housing is "five over one"—five stories of wood frame over a concrete ground floor.

In Europe, there never was a culture of building fast and cheap, and light wood frame was never accepted. Europeans are often shocked to find that one can kick a hole or punch their fist through most American walls. Most Europeans lived in apartments with solid clay tile walls, and if they could afford a house, they wanted it to last for generations. Wood might be used for floors and interior walls, but the exteriors were solid load-bearing masonry. The forestry industry was sophisticated and old: "The link between forest production and timber consumption by industrial societies was clearly laid out in 1713 by a Lower Saxony mining administrator called von Carlowitz who gave us the word *Nachhaltigkeit* or sustainability."[9] There are no virgin or old-growth forests in Europe; it is all managed.

Given the dislike of light wood framing, there was lots of research into how to use small pieces of wood more efficiently by turning them into bigger pieces of wood.

There is nothing new about mass timber; many old warehouses in North America were built with heavy timber columns and beams, with lumber nailed together on site for floors, what is now called nail laminated timber, or NLT. In the 1970s, Julius Natterer developed Brettstapel, a method of connecting low-grade timber together without glue. Natterer used kiln-dried hardwood dowels hammered into softwood timber; as the dowels absorbed moisture, they would expand, locking the wood together. This is now known as dowel laminated timber (DLT). The largest producer is StructureCraft in Abbotsford, British Columbia.

But the real mass timber revolutions started in Austria in the early 1990s with Gerhard Schickhofer, one of a group of engineers studying laminated wood. Unlike Brettstapel, or NLT, where the wood is all parallel, Schickhofer laid up his wood crosswise on top of each other. This could use smaller pieces of wood—our classic two-by-four—but was more stable dimensionally and could replace a two-way slab on columns instead of having to sit on beams. It could use waste wood too small for other purposes that could be spliced together. By 1998, the material, now known as Cross-Laminated Timber (CLT) was approved for construction by the Austrian government, and companies such as KLH started cranking the stuff out and many small buildings were built out of it.

In 2007, architects Andrew Waugh and Anthony Thistleton convinced a client to let them use CLT in a small apartment building in Hackney, then a not-so-prime part of London. The dubious client agreed but insisted that all the wood be covered with drywall so that none of the tenants would know; he was afraid nobody would want to rent if they did. The result was the first tall urban housing project made entirely of prefabricated CLT panels. The architects write: "Completed within 49 weeks, and delivering 29 fully insulated and soundproof apartments, the project demonstrated for the first time that CLT has the potential to be a financially viable, environmentally sustainable and beautiful replacement for concrete and steel in high-density housing."

The making of cement and steel are responsible for 15 percent of global CO_2 emissions, so when the industry realized that these products could be replaced with mass timber, it took off; every month, there seems to be a new plyscraper claiming to be the tallest wood structure.

Wood is 50 percent carbon by weight, taken out of CO_2 in the atmosphere and combined with hydrogen from water to make cellulose and lignin, respectively, the mass and reinforcing naturally found in wood. The oxygen from the CO_2 is released back into the atmosphere.

Many in the industry claim that the wood was carbon-negative (or carbon-positive, depending on who's talking), storing carbon for the life of the building. It's true that it is storing carbon, but calling it "carbon-negative" is controversial.

One builder of a Swedish project claimed in a presentation I attended that all the wood in the building was replaced by new growth in Swedish forests in forty-four seconds. More recently, the developers of what at the time of writing is the tallest mass timber structure in the world, a glass-clad residential tower in Milwaukee, claim that the wood is "replaced" by natural growth in North American forest in less than twenty-five minutes. But that means nothing when a forest is clear-cut.

Carbon savings with wood construction are counted in two

ways: **avoided emissions**, when compared what might have been emitted if building out of concrete or steel, and **carbon storage**, the carbon locked away in the wood. I have reservations about both; I recently reviewed a project where the architect used massive columns and beams in a design optimized for flexibility rather than fiber and credited the building with avoided emissions from not using concrete, and then called the project "carbon positive." How does that work?

I have always thought that counting avoided emissions was silly, very much like being on a diet and counting the calories of the chocolate cake I didn't eat instead of counting only the calories that I do. Paula Melton of BuildingGreen agreed, saying it reminded her of the old joke where a person goes shopping and buys a new sweater on sale. They come home and tell their spouse, "I just saved ten dollars on this sweater!" and the spouse says, "Show me the ten dollars."

Others disagree, with Kirksey Architects writing that "studies accounting for long-term carbon dynamics of wood products shows that the substitution effect of avoiding fossil fuel emissions is even more significant than carbon stored in wood."[10]

Carbon storage is also problematic; according to the Mass Timber Institute, "It is estimated that one cubic meter of mass timber sequesters one metric ton of carbon dioxide."[11] Some have said this is so wonderful that we should use more wood and store more carbon! But as British engineer Will Hawkins notes, "accounting for sequestered carbon is often a source of debate, confusion and inconsistency. When sequestration is reported as a negative emission, it can create the counterintuitive impression that using timber excessively can have environmental benefits." Hawkins notes that "carbon uptake in newly planted saplings is initially slow, but then accelerates as these become established. Just harvesting, processing, and constructing a timber building still results in a spike or burp of carbon, whereas sequestration occurs gradually." And, "Carbon accounting should always start at zero—credit should not be taken for a tree planted 50 years ago, even if this eventually ends up being used to build the structure under investigation."[12] The wood doesn't grow back in minutes or seconds, but will take fifty years until we know that the trees that were cut have been replaced.

Kirksey architects and others claim that mass timber production helps the forest.

> Using wood instead of other materials like steel or concrete helps create healthier forests. Using small pieces of lumber for the floors and beams instead of large timber pieces allows manufacturers to leave large trees in the forest. Large trees are better at absorbing carbon from the atmosphere and can help reduce fires and allow the remaining trees to flourish.

Lever Architects say much the same thing about sustainable forestry.

> It has been shown that forests that are treated as valuable ecosystems—that are managed for multiple benefits in addition to fiber production—have the capacity to improve ecological diversity, foster climate change resilience, reduce catastrophic wildfire risk, while at the same time sustaining the economies of rural communities that have depended on these forests for generations.

Chris Magwood of Builders for Climate Action wrote in a report on emissions from residential construction and declines to attribute any storage of carbon to timber construction.[13]

> There remain important and unresolved concerns with current accounting methods related to virgin forest products like lumber. Some of these concerns include uncertainty about the amount of carbon released from soils during logging operations; the amount of carbon returning to the atmosphere from roots, slash and mill waste; the amount of carbon storage capacity lost when a growing tree is harvested; and the lag time for newly

planted trees to begin absorbing significant amounts of atmospheric carbon dioxide.

Andy Simmonds and journalist Lenny Antonelli write that we should use wood more carefully.[14]

> In Sweden, natural forests are being systematically clear-cut and replaced with even-aged plantations, according to five environmental NGOs. Recent logging of protected forests has also been reported in Estonia, Lithuania, and Romania. And with extractive demand on land increasing, nature is more at risk of being squeezed out. So, while supporting a move from concrete and steel to timber and other natural fibres, our primary goal should be to dramatically reduce the quantity of raw materials needed in the first place. When specifying timber or other natural materials, how efficiently they are used can minimise the pressure on landscapes. As well as prioritising reclaimed or recycled materials where possible, smart choice of build system matters too.

It is all very confusing; I have been talking about wood construction for a decade and have not yet got a definitive answer about how much wood is stored in mass timber structures. A few years ago, Dave Atkins, an author of the Mass Timber report, told me that "the consensus in the research is that 50 percent of the carbon in the form of wood makes it to the mass timber." Some wood is left in the forest specifically to rot and provide animal habitat; some scraps are burned to kiln-dry the wood. But if the trees were left in the forest, fully 100 percent would eventually be released into the air, so 50 percent is pretty good. Atkins also notes that "if you don't grow it, you mine it."

Years ago, when I challenged architect Andrew Waugh with the 50 percent number, his flip answer was, "So we'll plant two trees!" But on a recent visit to his new Black and White building in London, he pointed out how slender the columns and beams were; they were made from laminated veneer lumber (LVL) that is peeled off logs with 90 percent efficiency. He proudly noted that their buildings now use 40 percent less fiber than when they started working with mass timber. Much has changed from when architects were proud of how much wood they used.

When I returned from my trip to London, I complained to Peter Moonen of Wood WORKS! Canada that I was no further ahead in understanding the true carbon footprint of mass timber. He said I never would; every forest is different, and there are 141 eco-types in Canada alone. Are the roots deep or shallow? What is the temperature or the humidity? "I think the one thing I can guarantee is that if you do a life cycle assessment is that it will be wrong."

All I can do is fall back on the call for **sufficiency**: How much do we need? Do we really need these silly wood towers, these "plyscrapers"? Or **materiality**: Is it the right stuff for every case? Andrew Waugh doesn't think so. He told me that twelve to fifteen stories are optimum for wood; if you go higher you run into trouble because it is so light. He was much more explicit in a recent quote to Dezeen: "It's bullshit, because if you're going to build a tall building in timber, you still have to fill it full of concrete to make sure it doesn't wave around."[15] Or **efficiency**: once wood buildings get tall, the columns get huge and eat up usable space. This is why steel took over from stone in the late nineteenth century. And finally, there is **frugality**: is this design achieving more with fewer resources? Is mass timber the best way to build with wood, or can other technologies do more with less?

British architect Piers Taylor has said, "Anything below two stories and housing isn't dense enough; anything much over five and it becomes too resource intensive." That's because mass timber uses four times as much fiber per square foot as does lightweight framing. That's why the two-by-four is so powerful and important; it lets you use less stuff, our prime directive here.

In Sweden, robots turn two-by-fours into multistory houses with quality and precision unheard of in North America. They have redesigned every connection and component for flexibility and repairability. The housing is solid, quiet, comfortable, resilient, and fast.

In the UK, Craig White is building social housing out of pre-fabricated panels filled with straw. He makes his houses light and moveable so that he can put them on land that might be waiting for development or approvals. He claims it is "less than zero carbon."

This is not all meant to be a criticism of mass timber! Every square meter of construction where wood replaces steel or concrete can be a plus for the climate, eliminating massive upfront emissions. It is instead a plea to use as little fiber as one can get away with when designing a safe and strong structure.

Perhaps the most pernicious story ever written was "The Three Little Pigs," with its very English conclusion that houses should be made of brick. The future we want is built of sticks and straw.

Insight: build with sunshine and sky, using less stuff.

The Block of Flats

In March 2023, a study titled "What Really Matters in Multi-Storey Design? A Simultaneous Sensitivity Study of Embodied Carbon, Construction Cost, and Operational Energy" was published in *Applied Energy* journal.[16] Hannes Gauch and his team constructed a magic box of a building model with many variables to find the optimal building with the lowest embodied carbon and operating emissions. They had knobs for the building's shape, size, layout, structure, ventilation, windows, insulation, air, and use for residential and office multi-story buildings across different climates. They also wanted to answer a question that many are discussing as buildings reduce their operating emissions, they increase their upfront or embodied carbon emissions:

> Embodied emissions in modern buildings are not only increasing relative to operational emissions but also

in absolute terms. This trend raises the question: Do significant trade-offs between embodied and operational emissions exist in building design?

Twisting all these knobs and variables, the magic box described what buildings want to be to minimize their upfront and operating emissions.

Buildings want to be wood. Mass timber comes in with the lowest upfront carbon, but not the lowest cost.

Buildings want to be short. The Goldilocks spot for upfront carbon is four to six stories. There are practical limits to how high you can build in mass timber, and six stories is usually the limit for light wood framing that's common in North America and Scandinavia. This model doesn't include light framing; the authors are from the UK and told me, "We have not included light timber frames in our study. It is not common in the UK to build larger buildings that way, but a comparison might be an interesting study to do!" But everything gets more expensive when you go tall, not just structure. Plumbing, structure, elevators, everything except land cost per unit, which is why developers want to keep piling on the floors. This is why I keep saying that the single biggest factor in the carbon footprint in our cities isn't the amount of insulation in our walls, it's the zoning; we get these tall buildings on busy streets and industrial areas because we protect vast zones of single-family housing.

Buildings want to be boxy. The study finds that compactness cuts heating and cooling in half and reduces upfront carbon and construction costs. This flies in the face of contemporary practice, where architects use computers and parametric design to increase complexity and surface area. In our homes, we see gables and jogs and bump-outs, when what we should be building is as close to a cube as possible.

Buildings want to have more wall than window. This is a problem for architects, who design windows more for their aesthetics than for their usefulness. Gauch writes in the study:

Decisions concerning windows are most influential for heating and cooling loads, especially the window-to-wall ratio. Whilst higher window-to-wall ratios decrease all three efficiency metrics, windows with lower U-values (triple and quadruple glazing) entail higher costs. This suggests a non-negligible trade-off between energy efficiency and construction costs.

This is one of the most interesting and possibly controversial findings. Architects and designers used to try to use windows as a source of free solar energy. Frank Lloyd Wright designed the Hemicycle House for the Jacobs family with a big curved floor-to-ceiling glass wall, which would allow the sun to warm the thermal mass of the stone and concrete floor. The family ended up getting dressed in the bathroom all winter because it was the only room with a radiator. "Mass and glass" was all the rage in the seventies when many architects and designers were experimenting with solar homes.

Meanwhile, in Saskatchewan, Harold Orr and Robert Dumont were asked to design a solar home in a place with cold winters and very short days. They came up with a design with super-insulated walls, heat recovery ventilation, and small windows. Essentially, it was the prototype for the Passive House or Passivhaus. It was generally ignored because, as Green Building Advisor's Martin Holladay notes, all these ideas were coming from "hippies and Canadians." Holladay, looking back at the thinking of the seventies, wrote in 2015:

> It turns out that every extra square foot of glazing beyond what is needed "to meet the functional and aesthetic needs of the building" is money down the drain. In a way, this advice is liberating: it compels the designer, secure in the knowledge that no technical or functional issues are at play, to think about aesthetic issues—and that's almost always a good thing.[17]

From today's viewpoint, it is not a good thing at all. Windows today are very good; I used to write that for heat loss, the best window is worse than a bad wall. Now you can buy windows that are thermally no worse than a mediocre or even decent wall. But they can have massive upfront carbon emissions. Working at the Daniels School of Architecture in Toronto, Kelly Alvarez Doran and Ted Kesik compared double-glazed windows to the much more efficient triple-glazed windows in six cities, noting that the latter had 50 percent more glass, 100 percent more spacers and 30 percent more frame. Alvarez Doran says, "The upfront emissions are the same in each city, but the operational emissions are variable based on grid intensity and degree days." It takes years for the carbon emissions saved by going triple-glazed to exceed the increased upfront emissions from making the windows—until 2040 in Calgary, and 2054 in Boston.

When architects design for aesthetic issues, the windows often get big; they are important design elements. Engineer Nick Grant tells architects to get a grip.

> Windows are much more expensive than walls and are lovely things, but truly a case of where you can have too much of a good thing, causing overheating in summer, heat loss in winter, reduced privacy, less space for storage and furniture and more glass to clean.

You still can read recent studies that claim, "Replace artificial light with daylighting and use lighting sensors to avoid demand for lumens from artificial light."[18] I suspect that with LEDs being so efficient, there is probably more energy lost through the windows than there is running the lighting.

The Gauch study concluded that "the results show that a building's shape and its window configurations are the most impactful design decisions determining operational heating and cooling loads for residential multi-storey buildings."

Windows are hard. There is a lot of talk these days about how windowless bedrooms make housing more affordable, but we need views and emergency exits. We need exposure to the sky for our bodies' circadian rhythms to sync with daylight. We need to see green for our biophilic response. But these views can be framed like a picture rather than a giant glass wall. I concluded years ago that we should "Keep the windows as small as you can get away with and still let in the light and views that you want, with an eye for proportion and scale. And keep it simple." But perhaps we need a new aesthetic where small windows are acceptable.

In the end, according to this model, the greenest building will be not too tall, made of wood, and a boxy simple form with windows designed to frame a view, not to make a statement.

Instead, we get the opposite. We get "Bjarked!", my term for buildings designed by Danish superstar architect Bjarke Ingels, whose buildings in Vancouver, Calgary, and Toronto all have complex forms, terraces on top of living spaces in almost every unit, vast, complex surface areas insulated with expensive vacuum panels, and are monuments to upfront carbon.

Professors Jo Richardson and David Coley have written that we have to change the way we think about buildings.[19] Less Bjarke and more boring. "If an architect starts by drawing a large window for example, then the energy loss from it might well be so great that any amount of insulation elsewhere can't offset it. Architects don't often welcome this intrusion of physics into the world of art." They call for "a revolution in what architects currently consider acceptable for how houses should look and feel. That's a tall order—but decarbonising each component of society will take nothing short of a revolution." I wrote in one of my complaints about Bjarke: "If we are going to ever get a handle on our CO_2, we are going to see a lot more urban buildings without big windows, without bumps and jogs. Perhaps we might even have to reassess our standards of beauty."

One of the most interesting blocks of apartments I have ever seen was the R50 cohousing project in Berlin, designed by Jesko Fezer with Heide & von Beckerath Architects. It is a simple six-story block, clad in wood, with a clipped-on steel balcony with chain-link fencing for the guard. I have no doubt that many would call it ugly. It's owned by the occupants in a form of tenure called Baugruppen, or building group, a method of building architect-led, collectively funded, community-based living. The city of Berlin offered sites to groups instead of selling them to conventional developers, and Fezer took the lead, telling *Metropolis*: "We have to think about the role of the architect if we have a social idea of how the profession can contribute to society."[20] Working on a budget means less upfront carbon, as all the fancy finishes of a conventional condo are omitted here. You see the concrete ceiling and floor; nothing is more minimal than chain-link fencing. Interiors are completely flexible. Architect Michael Eliason writes:

> Baugruppen allow for an environment to answer questions that market-rate development cannot. What could housing specifically designed for single parents or co-parents look like? Is it possible to design homes for multiple generations living in one unit? What about housing for cohabitating families? I have never seen market rate development feature space that could be used as a teenager's suite—but I have seen dozens of Baugruppen that incorporate this. Projects with smaller bedrooms for children are also common—this has the advantage of reducing unit size and cost.[21]

Our North American model of the development process starts with expensive land, high development fees, and expensive, very tall construction, with developers competing with fancy finishes and architectural pizzazz but little substance, pumping up the upfront carbon.

We need a radical rethink of how and what we build for a climate crisis. We need the right kind of zoning that permits little apartment buildings everywhere. We need the height done right. We need more flexible designs, with single stairs permitted in lower buildings. We need simpler, boxier buildings made with low-carbon materials such as wood and straw and old newspapers rather than steel and foamed plastic. We need more flexible forms of tenure that allow for greater flexibility and shared resources. We need every building to be designed to the Passivhaus standard. And, of course, we need them to be in fifteen-minute cities so we don't need parking or cars.

None of this requires new materials or building technologies; we could do it tomorrow. It's that zoning thing again, and political will.

Insight: The best place to live with the lowest carbon footprint is a relatively small apartment in a small building in a walkable community.

The Folly of Foam Insulation

A dozen years ago, almost every green building was full of foam. It was wonderful stuff; it had the highest insulating value per inch and made a tight seal that almost eliminated air infiltration. I have spray foam insulation above my head in an addition built in 2014 by the greenest construction company in the city.

The story of how foam insulation went from being a go-to solution to a pariah in the green building industry is the story of upfront carbon, the story of how priorities have begun to change.

A dozen years ago, you couldn't have too much foam in your home. I have been in houses designed for maximum energy efficiency that had eight inches of polyurethane foam, sealing it up tight as a drum and reducing heat loss and air infiltration to almost zero. How green is that? Some companies offered an even greener foam made with soya oil instead of fossil fuels. If you were worried about energy efficiency, foam couldn't be beat.

Back in 2013, all anyone seemed to care about was energy efficiency. But discussions about carbon emissions were in the air, and builder and author Chris Magwood began looking at the carbon content of building materials. He writes:

I called this the carbon elephant in the room because in 2013, hardly anybody was accounting for this significant amount of emissions. Instead, all the focus was on making buildings more energy efficient. While efficiency is important, reductions in material emissions are immediate and therefore more impactful in reducing atmospheric carbon concentration now, rather than in the future.

I saw Chris do a presentation of his findings in 2015 and was floored; he demonstrated that an energy-efficient home insulated with polyurethane foam had total lifetime carbon emissions that were greater than a home built to the crappy Ontario Building Code because of the fossil fuels the foam is made of, but more importantly, the blowing agents that made it foamy.

One can't overestimate the importance of Magwood's findings; a generation of builders and architects all went, "OMG, we have been doing everything wrong!" It was like a light switch being flipped. Here was a self-trained straw-bale builder in Peterborough, Ontario, who changed the industry. The manufacturers have been coming up with better blowing agents with lower global warming potential, but Magwood says "they're still half as bad" because the foam itself is still made from fossil fuels. Builders have been switching to natural insulations such as cellulose and wood fiber, or plastic-free insulations like glass and rock wool, but when it comes to stopping air leakage and heat transfer, they are half as good.

Looking in the rearview mirror, it is surprising that we were so fond of these products, but they really were magical, giving us high insulation values, air sealing, and even a bit of structural reinforcement in one product. But the high upfront carbon was only the final straw that broke the proverbial camel's back.

- Many of the chemicals used to make foam insulation are toxic, and some people developed serious chemical sensitivities to it if it wasn't properly installed.

- It was extremely flammable—one insurance expert called it "solid gasoline," and it released hazardous chemicals when it burned. They were loaded up with brominated flame retardants, which are "persistent, bioaccumulative, and toxic."
- It was sticky and hard to remove, making renovation or disassembly more difficult

Surprisingly, it is not going away. An industry rep says, "The major growth drivers for this market are the stringent government regulations for greenhouse gas emissions and the increased demand for energy efficiency in homes and buildings." He extolls the virtues of the new blowing agents: "With the future in full sight and the evolution and adoption of HFO blowing agents in more SPF products, green builders, architects and contractors alike see great potential for their building projects, knowing that air pollution issues and environmental considerations have been heard."

There is no question that foam insulation is better than it was, now that the blowing agents have lower global warming potentials and do not destroy the ozone layer. But to claim that "environmental considerations have been heard" and dealt with is a bit of a stretch. It still ignores the danger of fires, toxic fumes, outgassing, and the loading of toxic brominated flame retardants. But there are so many vested interests: the fossil fuel companies, the blowing agent companies, the bromine industry, but most importantly, the construction industry that wants a quick and easy way to seal and insulate.

Replacing it requires layers of different materials for controlling moisture and air movement and a lot more care and precision during installation. This sounds too much like work. It's also expensive. But Chris Magwood's discovery of the carbon elephant in the room changed the green building industry by exposing the folly of foam.

Insight: Avoid plastic foam insulation; it is a toxic solid fossil fuel.

The Heat Pump

There are two innovations in electrification that I believe are worldchanging: the e-bike and the heat pump. There is nothing new about either of them, but incremental improvements have brought them to a tipping point.

Heat pumps have been in our houses for a century; your refrigerator is one. It has an evaporator side where a compressed refrigerant turns from liquid to gas, a process which absorbs heat, keeping your food cool. The gas is then compressed into a liquid, which releases heat outside of the fridge. An air conditioner does this too, removing heat from your home and dumping it outside. Turn the air conditioner around so that it removes heat from the outside air and moves it inside, or install a valve so that it can work both ways, and you can call it a heat pump.

Twenty years ago, most heat pump installations were "ground source," with pipes drilled into the ground and filled with liquid. Everyone called it "geothermal heating" and claimed it was renewable, taking heat from the earth or the sun that heated the earth. Calling them geothermal was a lie; it was a heat pump, moving heat from the ground when heating, and dumping it into the ground when cooling. True geothermal heating is found in Iceland and Japan, and may well become common with extreme deep drilling in a few years. These systems, properly installed, worked very well; the problem was that all that drilling and piping was very expensive, and you never know what you are going to hit when you drill, so pricing was always a gamble. The costs and the limitations, along with a lot of sketchy installers, limited their use.

Air source heat pumps (ASHP) have been around for decades, but were used mostly in warmer climates that needed a bit of heat occasionally. Their efficiency would drop significantly when the outside air is colder, and additional resistance heaters were required in northern climates. Consumers distrusted them and were often told that there wasn't enough heat in the Canadian air. This isn't true; there is lots of heat in air that might be –10 degrees Celsius, but that's still 263 degrees Kelvin. With the development of inverter-based compressors and other refinements, ASHPs now work well in seriously negative temperatures.

Where a conventional electric heater makes heat through resistance at close to 100 percent efficiency, that is still way more expensive than gas. Heat pumps don't make heat but move it, which is even more efficient, averaging 300 or 400 percent, which is competitive with or cheaper than heating with gas, and you get an air conditioner thrown in. They are not as efficient as a ground source system; the earth is a better heat sink and stays at about the same temperature all year, but are far cheaper and can be installed anywhere. Finally, an affordable and effective way to heat without fossil fuels! Everyone is excited. Journalist David Roberts cheers, "Fist pumps for heat pumps!" TED talker Saul Griffith says, "Electrify everything!" Joe Biden throws serious money at them.

There's always a catch.

The biggest problem is that refrigerants are greenhouse gases, and heat pumps leak throughout their lives, but also during installation and maintenance. It averages out at about 5 percent per year.

The Montreal Protocol started the process of getting rid of the worst ozone layer destroying chlorofluorocarbons (CFC) such as Freon, and the more recent Kigali amendment promises to get rid of the worst hydrofluorocarbons (HFC), but even the latest and best versions of HFCs have global warming potentials (GWP) of 675 times that of carbon dioxide.

Some refrigerants have very low GWP; most new fridges are filled with butane, but fridges don't need much refrigerant. Carbon dioxide has the lowest GWP, but CO_2 heat pumps can only heat, not cool. Propane also works well, but it is obviously flammable, and the amounts permitted are limited. The other problem with it is that it is cheap and available anywhere, whereas there is big money for Chemours and Honeywell in making fancy proprietary HFCs. So, they screamed, "FIRE!" They delayed approval of propane (known as R290), even though it has been approved in Europe for years and the maximum permitted weight, a kilogram, or 2.2 pounds, is a tenth of what everyone has in their barbecue tanks. In Europe, you see all kinds of R290 heat pumps; in North America, you can barely find them.

You will hear a lot about the greenhouse gas problem from HFCs and the fire danger from R290 in the next few years as fossil fuel interests try to hang on to home heating. Still, the HFC leakage problem, while serious, is minimal compared to the CO_2 released from burning gas, and complaining about 2.2 pounds of propane outside your home compared to piping an explosive gas into your furnace and water heater is just ridiculous. They will talk about how you need a second source of heat for resilience; what if the electricity goes out? They don't mention that furnaces and modern water heaters all have electrical connections for fans and pumps.

The solution to the leakage problem is twofold: use a "monobloc" design where all the refrigerant is in the outside unit and a separate water loop brings the heat or coolth inside. In colder climates, this adds a bit of cost for antifreeze. But with monoblocs, installers do not need special refrigerant training; any plumber can do it because the refrigerant is sealed inside the unit.

The second, more difficult step, is to design our homes around the maximum capacity of an R290 heat pump. This will require better insulation and windows, and the bigger the house, the higher the standard. This will also address resilience; a well-insulated house stays warm for hours or days when the power goes out.

"Fist pumps for heat pumps!" has a nice ring to it. "Fist pumps for R290 monoblocs!" doesn't roll off the tongue so nicely. But this is what we need for the heat pump revolution.

Insight: The world is moving to heat pumps, but keep it small and ask what the refrigerant is.

The Puffer Jacket

The puffer jacket is the epitome of ephemerality; there is nothing to it, a bit of parachute fabric and down or light synthetic insulation. It weighs almost nothing and squeezes down to fit into your pocket. It's a fascinating story of how smart, minimalist design can do a better job with less stuff.

The puffer was invented by George Finch, a brilliant Australian chemist and mountain climber, considered one of the two best in the world, along with George Mallory. The two were part of a 1922 expedition to Mount Everest, where Finch showed up in what he called his eiderdown coat, made of bright green hot air balloon fabric. The other climbers all wore cotton and tweed and considered Finch's outfit a joke. The expedition secretary wrote, "They have contrived the most wonderful apparatus that will make you die of laughing." One can imagine him taking the abuse and quoting the most famous line from his son Peter Finch's film career: "I'm mad as hell, and I'm not going to take it anymore!" Instead, he outlasted the skeptics, writing later: "Everybody now envying…my eiderdown coat, and it is no longer laughed at."

In 1936, Eddie Bauer developed the Skyliner quilted down jacket after almost freezing to death on a fishing trip. At 21 ounces, it was marketed as "lighter than feathers, warmer than ten sweaters." Of course, if you look at most American websites, you will be told that Eddie Bauer invented the puffer jacket.

Today, puffer jackets are everywhere. They are what *Financial Times* writer Robert Armstrong calls "Gorpcore."

> Gorpcore—functional outdoor clothing worn for everyday rather than to stay warm and dry in the elements—turns out to have an alpinist's endurance. It was first noticed in the wild six or seven years ago, its ascent roughly coinciding with camping-chic runway shows from Prada and others. Puffer coats, Patagonia fleeces and technical-looking shoes have been seen in New York bars ever since.

When I first read Armstrong's quote there, I didn't understand it. People wear functional clothing every day because it keeps them warm and dry in the elements. I have been wearing it for decades. Freud supposedly said, "Sometimes a cigar is just a cigar," but evidently a puffer is more than just a puffer.

Morweena Ferrier of the *Guardian* suggests that it is environmental signalling:

> There is something in flagging your allegiance to clothes traditionally worn outdoors. It is not simply that hiking and camping have a virtuous reputation—you hike, therefore you care about the environment. If you ski, you are probably rich. It is a coat favoured by winter-sport enthusiasts, who tend to be affluent. The function attracts the wealthy, who imbue the puffer with lifestyle status.

I still have my first Mountain Equipment Co-op fleece top from when they opened in Toronto in 1985. I am typing this while wearing my rowing tights. Of course, they are black; I trained as an architect. Some say that architects wear black for purely functional reasons; when we drafted with pencils, our hands and sleeves would rub across our drawings and smudge our clothing. I used to be covered in ink from my Rapidograph pens. But almost nobody has used pencils or pens for thirty years. Cordula Rau asked one hundred architects for her 2008 book, "Why do architects wear black?" and got many responses, including "Because they believe they need to stand out from the bourgeois society whence they come if they are to be real 'artists'" Other reasons include French architect Eduard François', who said simply, "Architects wear black because they are sad." I have met François, and he most definitely was not.

When I look in the mirror, I see functional, black clothing. My puffer is black. Even my rain gear, which should be fluorescent yellow or orange for safety, is black (this was a mistake.) All of which is to say that even for someone like me who adores the practicality and ephemerality of the puffer jacket and gorp-core, I am still a slave to fashion.

I sometimes feel guilty about wearing a puffer filled with down instead of some of the alternatives such as polyester, but the totally disinterested International Down and Feather Bureau has a life cycle assessment prepared by a reputable third party, which concludes that down has "eighteen times less impact on climate change than polyester fill." Polyester is a solid fossil fuel, so this should not be surprising compared to a natural product. They also claimed that "on a per ton basis, down has an 85%–97% lower impact than polyester in all categories analyzed." I thought this might be silly to do a comparison on a "per ton" basis given how light down is, but they compared material with the same CLO value, a measure of how well clothing insulates. One hundred eight grams per square meter (0.3 ounces per square foot) of down has the same CLO as 230 grams (8.1 ounces) of polyester, so while the "per ton" comparison is not as bad, it is still relevant. As is common outside the construction industry, the life cycle assessment is thin and more of a summary, and it doesn't give hard numbers. But I feel less guilty about down from an environmental point of view.

Feeling guilty about down from an animal welfare point of view is another story. Some of it comes from geese raised for foie gras, which involves force-feeding to fatten the liver and is cruel. Others come from live-plucking, which is painful. Ninety percent of down comes from China, where duck is a common meal.

Patagonia has developed its own Advanced Global Traceable Down Standard (TDS) that other brands could sign up to. Alas, they appear to be on their own. "Though we continued to advocate for other brands to begin sourcing Advanced Global TDS–certified down, no major brand ever did, despite continued efforts."

Other companies adhere to a Responsible Down Standard (RDS) that certifies birds "have not been subjected to treatments that cause pain, suffering or stress and that an identification and traceability system that applies to the origin of the material is applied and maintained."[22] Companies also note:

- down, feathers, and hides must be by-products of the food industry
- the use of down and feathers obtained by live plucking of animals is prohibited
- the use of down and feathers from the foie gras industry is prohibited

PETA doesn't believe any of this and has exposés on their website of certified "responsible down suppliers" demonstrating animal cruelty. "The RDS is a veil of false assurances about animal welfare that do little or nothing to protect the animals who continue to be exploited and killed for profit."

This makes it all a very hard choice. From an upfront carbon point of view, down is clearly superior to polyester. It's also functionally better, at half the weight for the same amount of warmth and a lot more compressible when packing it away. And we haven't even gotten started on the problems of microplastics that are shed from polyester clothing products and what they are doing to the marine environment.

Patagonia is expensive. My UNICLO down jacket is filled with RDS-certified down. Vegans and PETA say we shouldn't be benefiting from the slaughter of animals. But this seems to be one of those cases where from an environmental and especially an upfront carbon point of view, responsible down wins hands down.

Then there is the fabric that encloses the down. Almost all puffers are made with polyester fabric. Patagonia uses recycled polyester and has started using "prevented ocean plastic," which is collected from coastlines that lack waste management infrastructure. "For the Spring 2023 season, 87% of our polyester fabrics are made with recycled polyester. As a result of not using virgin polyester, we avoided emitting more than 4.4 million pounds of CO_2e into the atmosphere."

Many other companies claim to use recycled polyester from

PET bottles, but there's a catch; according to one study, making polyester from recycled plastic has ten times the carbon footprint of virgin polyester. "The total CF of recycling processes was much larger than that of virgin production processes. This was caused by a series of energy intensive procedures (e.g., crushing, high temperature cleaning and drying) involved in the spinning stage from waste polyester bottles to recycled polyester fibers."[23]

Wait, there's more. The Changing Markets Foundation claims that recycling polyester from bottles into fabric is not sustainable or circular because "PET bottles should be kept in a closed-loop recycling system for food contact materials." The only truly circular polyester strategy is "fiber to fiber" where it is made from clothing; otherwise, the PET bottle companies have to buy virgin plastic to replace what was turned into clothing.

The puffer jacket wraps up so many of the issues of sustainability in one warm coat. From a design efficiency and ephemerality point of view, it is brilliant, doing so much, so well, with so little. From an information point of view, we have so many companies saying they are measuring their impact and so little real hard data for a product with only three components (fill, jacket, zipper). Only Allbirds tells us what we want to know: 20.9 kilograms (46 pounds) CO_2e emissions. We have to weigh animal welfare against embodied carbon against marine environment welfare. We even have to decide whether we like recycled water bottles or demand recycled fishing nets. It is all so complex, all about a relatively simple product.

Insight: In the end, we come up with the same answer every time, whether it is my coat or my computer:

1. Buy nothing and make do with what you have.
2. Buy secondhand.
3. Buy high-quality and make it last. (Hello, Patagonia!)

The Hamburger

In 2019, presidential advisor Sebastian Gorka complained about Democrats and their proposed Green New Deal: "They want to take your pickup truck. They want to rebuild your home. They want to take away your hamburgers. This is what Stalin dreamt about but never achieved."

Ignoring the fact that Stalin actually liked hamburgers and introduced them to Russia, Gorka seized on two icons of American life: the pickup truck and the hamburger. They go together because the story of the hamburger is a story of mobility. Like the disposable coffee cup, the boundaries of the carbon footprint of a burger go way beyond the meat itself.

But let's start with the meat itself, with a McDonald's quarter pounder, because it is a straightforward quarter pound of hamburger, precooked. That's 113 grams (4 ounces). According to Our World in Data, a kilogram of beef has greenhouse gas emissions in carbon dioxide equivalents of 99.48 kilograms (219.3 pounds), giving the meat in our burger a carbon footprint of 11.24 kilograms (24.8 pounds).[24] Other sites put everywhere from 2.99 kilograms (6.6 pounds)[25] to 9.73 kilograms (21.5 pounds),[26] but I tend to trust the OWID gang.

Then some say that we shouldn't count the methane from cows since it has a shorter lifespan than CO_2 in the atmosphere, and others who say that the methane is biogenic since it comes from eating grass. Even if you accept either of these, the upfront carbon from the hamburger is still 41 kilograms (90 pounds) of CO_2e per kilo/pound. An American researcher, Frank Mitloehner, claims that American beef has a far lower carbon footprint at only 22 kilograms (48.5 pounds) of CO_2e per kilo/pound.[27] He's complaining about another professor, Tim Searchinger at the World Resources Institute, who calculates that beef has a footprint of 188 kilograms (414.5 pounds) of CO_2e per kilo/pound. The difference is mostly in land use; much of the carbon footprint of beef is attributed to the continuing destruction of forests to increase grazing lands. Mitloehner says this doesn't happen in North America; Searchinger says meat is a global product, and you can't look at the United States in isolation. "An increase in U.S. beef consumption, for example, can result in deforestation to make way for pastureland in Latin America. Conversely, a decrease in U.S. beef consumption can avoid deforestation and land-use-change emissions abroad."

Whew. This has been a consistent problem throughout this exercise; as soon as you get away from building materials, the numbers can be wildly inconsistent. Fortunately, the disinterested nerds at Our World in Data come up smack in the middle of the duelling professors.

Americans buy approximately fifty billion burgers a year, averaging three per week, with sales of burgers totalling $92.2 billion in 2021.[28] Just cutting back on burgers could reduce their impact on climate significantly. Tim Searchinger and associates at the World Resources Institute write:

> Reining in climate change won't require everyone to become vegetarian or vegan, or even to stop eating beef. If ruminant meat consumption in high-consuming countries declined to about 50 calories a day, or 1.5 burgers per person per week—about half of current U.S. levels and 25% below current European levels, but still well above the national average for most countries—it would nearly eliminate the need for additional agricultural expansion and associated deforestation.

But once again, we are talking about boundaries. The impact of the hamburger goes far beyond just the meat. So how did they become so big, such a dominant part of our culture? Why are they held up as American icons along with the pickup truck? Why do we eat so many of them?

According to Andrew F. Smith in his book *Hamburger: A Global History*, the Hamburg steak first got wide publicity in the United States at the Philadelphia Centennial Exhibition in 1876. It was eaten on a plate with a knife and fork. The first good meat grinders were also shown at the exhibition, and butchers and entrepreneurs quickly put two and two together. Smith writes:

The meat grinder was a great asset to butchers, who could now use unsaleable or undesirable scraps and organ meats that might otherwise have been tossed out. It also became possible to add non-meat ingredients to the ground beef, and it was very hard for the consumer to know what was actually in the mixture. Ground meat was cheap, ideal to sell to the working classes, and by adding even cheaper fillers, such as gristle, skin and excess fat, the butcher could enhance his already substantial profit.

The Hamburg steak became a hamburger sandwich when workers needed lunch at the new factories and mills, and lunch wagons and carts opened to serve them. Smith notes: "Because many customers ate their food standing up, placing the beef patty in a bun made sense. Who was the first lunch wagon proprietor to sell the Hamburg steak between two pieces of bread is unknown, but, by the 1890s, it had already become an American classic."

This is the key inflection, the turning point: putting it between two pieces of bread made it portable and fast. You couldn't eat a hamburger patty with your hands, but you could when it became a hamburger sandwich.

The first known instance of the concept of the sandwich is in Jewish tradition, in the story of Passover, which might well be the story of the invention of fast food. I am writing this on the first day of the Jewish holiday when Jews eat matzohs, which was bread made in a hurry; everyone had to pack up and leave Egypt, and the bread they were baking didn't have time to leaven. In the first century BC, Rabbi Hillel made a sandwich from the Passover plate, and Jews recite: "This is what Hillel did, at the time that the Temple stood. He wrapped up some Pesach lamb, some matzah and some bitter herbs and ate them together"—the first known written description of takeout food for people to eat while standing or walking, or in this case, in a hurry to get out of town.[29]

The first recorded use of the word in English comes from the historian Edward Gibbon, who wrote in 1762 about seeing "twenty or thirty of the first men in the kingdom...supping at little tables, upon a bit of cold meat, or a Sandwich." The eponymous earl credited with inventing it in England had traveled in the eastern Mediterranean, where stuffed pitas (falafel, anyone?) were common.

Then, as now, the sandwich was all about being fast and portable food that could be eaten without plates or cutlery. But street meat had a bad reputation; Smith notes that the term "hot dog" came about as a derogatory term for sausages in a bun, which some thought were adulterated with dog meat. Pure food activists were appalled with the hamburger sandwich because of the contents, with some complaining:

> The hamburger habit is just about as safe as walking
> in an orchard while the arsenic spray is being applied,
> and about as safe as getting your meat out of a garbage
> can standing in the hot sun. For, beyond all doubt the
> garbage can is where the chopped meat sold by most
> butchers belongs, as well as a large percentage of all the
> hamburger that goes into sandwiches.

Soon the street vendors and carts selling burgers were pushed off the road by the car, so they moved indoors to cafés and diners, many of whom sold burgers of dubious quality.

In 1916, Walt Anderson tried to change the hamburger for the better. He opened a shop with a window where customers could see him grind his own beef and make fresh patties in front of their eyes. He then partnered with an investor to open the White Castle, with white tile interiors representing purity and cleanliness. It was a huge success. They opened in urban areas near bus and trolley stops and close to large factories.

But things changed after the Second World War with the boom in car ownership. The inner-city hamburger chains didn't have parking, and everyone was moving to the suburbs, a wide-open market for new chains with new ideas. The McDonald brothers applied an assembly-line model to their California restaurant to deal with staff shortages, as well as switching to disposable cutlery and plates. They outsourced much of the work to their customers. Smith writes:

> This assembly-line system provided customers with fast, reliable and inexpensive food; in return, the McDonald brothers expected their customers to queue, pick up their own food, eat quickly, clean up their own rubbish and leave without delay, thereby making room for others.

This was the true start of the extravagantly wasteful linear food system we have today. The McDonalds refused to provide any indoor seating; there might be a few picnic tables, but customers were expected to eat in their cars.

Everyone showed up at their window to copy it; that's how we got Burger King, which started as a Florida knockoff. A man named Glen Bell went Mexican and called it Taco Bell. Today, much of the world's food is served on the McDonald's model.

That model means every 11.24 kilograms (24.8 pounds) of CO_2 from the burger leaves a trail of paper and plastic waste that is now the purchaser's responsibility. If bought at a drive-through, there is the idling time; a recent study found that this had increased to an average of six minutes and twenty-two seconds per customer.[30] According to Natural Resources Canada, a 3-litre engine idling releases 69 grams (2.4 ounces) of CO_2 per minute, so the average wait time adds almost another half a kilo of CO_2 to the meal. And, of course, we haven't included the drive to the restaurant in the first place, the drive that is required because of the design of our suburbs that are based on the automobile.

Smith concludes his history of the hamburger with the statement:

> During the past century, the hamburger sandwich has rapidly evolved, and it is likely to continue to do so in the future. Whether served in a restaurant, a fast-food outlet or the home, the hamburger is here to stay.

I am not so sure. As noted in our discussion of the coffee cup, we have to rethink whether speed and portability are such a good idea.

The World Resources Institute says we have to cut our consumption of hamburgers in half; even if that cuts the burger's carbon footprint in half because it takes out land use changes, it is still at almost 6 kilos, still a very big chunk of carbon. That's why so many others say we have to give up red meat altogether. Sebastian Gorka was not wrong; we do want to take away the pickup truck and the hamburger. We can't afford the upfront carbon emissions of either.

Insight: Walk to your nearest burger joint, and don't do it often.

The Car

The International Organization of Motor Vehicle Manufacturers is breathless in its descriptions of the wonders of the auto industry:

> Automobiles are a liberating technology for people around the world. The personal automobile allows people to live, work and play in ways that were unimaginable a century ago. Automobiles provide access to markets, to doctors, to jobs. Nearly every car trip ends with either an economic transaction or some other benefit to our quality of life.[31]

The American Alliance for automotive innovation is just as excited:

> A robust auto manufacturing sector is vital to a healthy U.S. economy. Autos drive America forward by supporting a total of 10.3 million American jobs, or about 8 percent of private-sector employment. Each job for an auto manufacturer in the United States creates nearly 11 other positions in industries across the economy.[32]

The Alliance claims that auto manufacturing drives $1.1 trillion into the economy and is 5.5 percent of the nation's Gross Domestic Product. These are vast numbers on their own, but they grossly understate the automobile's impact on our economies. In the United States alone, 37,471 people were killed and 2,443,000 injured in car crashes in 2016.[33] An NHTSA study released in 2014 estimated that the loss in productivity and lives cost $1.1 trillion per year.

Then there is the impact on climate, with 16 percent of global greenhouse gas emissions coming from the tailpipes of cars and trucks.[34] This will fall quickly as vehicles are electrified, but it doesn't address the problem of the upfront carbon emitted while building all those automobiles.

We don't know the exact upfront carbon of electric cars, but we can do a rough approximation from life cycle analyses prepared by Tesla and the International Council on Clean Transportation (ICCT).

The Tesla LCA is maddening. It tells us that the manufacturing process of a Model 3 is higher than that of an equivalent internal combustion engine-powered vehicle and that the Tesla will break even and have lower lifetime emissions after 5,340 miles, which is very low, much lower than electric car skeptics have used. It describes how the company is continuing to reduce the carbon footprint of its manufacturing. It tells us what is included: "raw and semi-finished material production including transportation, mechanical processing and shaping, battery manufacturing, vehicle assembly and paint shop, all fuels and energy (natural gas, electricity, etc.), other auxiliaries (lubricants, water, etc.) and end-of-life disposal."

It tells us what the boundaries are—where the calculations end—"exclusive of: capital goods (e.g., machinery, buildings), infrastructure (e.g., roads, power transmission systems), employee commute, external charging equipment and infrastructure, maintenance and service during use, packaging, transport to recycler, disposal of manufacturing waste, inbound transportation from Tier 1 suppliers, distribution to customers."

But then it never tells you what these numbers actually are, how much is from the batteries, and how much is from the car.

The best number we get is that the manufacturing phase of an average US Model 3 has a carbon footprint of 52 grams (1.8 ounces) per mile driven, and the average mileage of an American car is 200,000 miles, giving us a total upfront carbon emissions of 10.4 tonnes.

Going through the ICCT LCA is even more aggravating when you realize how these numbers seem to be picked out of the air, but at least they show a formula to get a rough number, which we can apply to different vehicles such as our nemesis, the Ford F-150 Lightning. It is surprisingly simplistic:

Gasoline (ICEVs) have upfront carbon emissions of
5.2 TCO$_2$e/ T of vehicle.
Battery-powered vehicles without the battery are
4.7 TCO$_2$/ T.

Producing your average American battery pumps out 57 kilograms (125.7 pounds) of CO_2 per kWh, 69 kg/kWh in China. That's it; that is the basis of the calculation of embodied carbon of the full life cycle, including recycling. We have no idea of what proportion is upfront, but will assume the great majority of it is. Running the numbers for an F-150 Lightning, we get 21.244 tonnes, slightly over twice that of the Model 3.

A small European or Chinese car is probably around 8 tonnes, so for this exercise, let's pick a worldwide average of 10 tonnes for easy math. In 2021, there were 80 million motor vehicles built worldwide, of which 21 million were built in China.[35] If we pretend they are all electric, that's 800 million tonnes of upfront carbon emissions, or 0.8 gigatonnes. Yet to stay under the 1.5-degree carbon budget ceiling, emissions must drop to 20 GT per year by 2030. If we plan to replace all of the 1.6 billion gasoline-powered cars on the road worldwide, we get upfront carbon emissions of 19 GT. The remaining carbon budget estimated at the start of 2023 was about 260 GT of CO_2, so making clean cars eats up 7 percent of the remaining budget. If they are all SUVs and pickups, it eats up 15 percent. This is why electrification may save the auto industry, but it doesn't save the climate.

This is a fundamental problem when it seems that the entire world depends on the auto industry. It's not just the vehicles, but the roads they drive on, the parking garages they sit in, the sprawling development they make possible, the hospitals they fill, and even the police state that evolved because of them.[36] The car isn't just a big part of the economy, but it often seems that it is the economy.

Of course, the ironclad rule of carbon applies to cars, and the manufacturers understand this. There are many ways they can cut the upfront carbon emissions of manufacturing: The Swedish electric car manufacturer Polestar is promising a climate-neutral car by 2030 using HYBRIT steel made with hydrogen and zero-carbon aluminum made with inert anodes.[37] A McKinsey study claims that the industry could reduce upfront carbon emissions by 66 percent at no extra cost by using recycled materials, green aluminum, and renewable energy.[38]

But coal-free aluminum and steel are decades away in the United States and China, and if the issue of upfront carbon is on the industry's radar, it is very far away.

The cars all have to drive on roads, which are often made of concrete. Even when they are topped with asphalt, the roadbeds underneath are made of cement-stabilized gravel. In China alone, infrastructure construction in 2016 pumped out about 250 million tonnes of CO_2.[39] An MIT study found that the annual emissions of construction materials used in the US pavement network were between 11.9 and 13.3 million tonnes per year, "equivalent to the emissions of a gasoline-powered passenger vehicle driving about 30 billion miles in a year."[40] The American Way is to pretend to reduce carbon emissions by adding lanes to highways with infrastructure investment, even though it is proven that more highways attract more vehicles through induced demand, and they rarely calculate the upfront emissions.[41, 42]

We Don't Just Need Electric Cars, We Need Fewer Cars

Readers may complain that I am barely mentioning the fact that in their full life cycle, electric cars have vastly lower carbon emissions than ICE cars. But this is a book about upfront carbon, not operating emissions. We simply do not have the carbon budget for all of this. Yes, we do need electric cars, but we also need a lot fewer cars and the infrastructure that supports them. As Matthew Lewis of California YIMBY notes:

> At this stage, if electric vehicles are to play a major role in solving the climate crisis—which they must—they have to be paired with dramatic land use reform that shortens or eliminates a substantial portion of all vehicle trips, and replaces them with transit, walking, biking, shared vehicles, and other forms of mobility. Only by combining a rapid deployment of electric vehicles with an equally rapid elimination of the need for most Americans to own and drive a personal vehicle in the first place can we have a shot at climate stability.[43]

The automobile manufacturers say, "The personal automobile allows people to live, work and play in ways that were unimaginable a century ago." This is true, but it is not necessarily a good thing. The manufacturers call cars a liberating technology, but it is the opposite, chaining us to a high-carbon lifestyle where we are dependent on the car for access to markets, to doctors, to jobs. As Lewis notes, we have to reimagine how we live now so that we don't have to depend on a car for every trip.

How Did They Get So Big and So Dominant in Our Culture?

The *Economist* explains why Americans have not been buying electric cars: "The American love affair with the internal combustion engine is an enduring passion, meaning that the road to electrification will be long and winding." The words sounded familiar; according to author Peter Norton, they were first used by Groucho Marx on October 22, 1961, on a TV show called "Merrily We Roll Along," subtitled "The Story of America's Love Affair with the Automobile." The show was sponsored by DuPont, which owned 23 percent of General Motors.

In his book, *Incomplete Streets*, Norton writes: "Motordom did not believe Americans loved cars enough to bring about the motor age unaided. Rather, it so feared the hostility to automobiles, especially in cities, that it organized perhaps the greatest private-sector public relations effort ever undertaken." They are still at it. Ads for pickups and SUVs wow us with power, speed, and manliness, encouraging us to want more. Car advertising should be banned, and while we are at it, so should SUVs and pickup trucks. At the very least, the industry should adopt a few of our strategies for sufficiency:

Frugality: This term was specifically applied to cars, which could all last longer, be a lot less complicated, and be a lot smaller. SUVs and urban pickup trucks are the antithesis of frugal.

Circularity: I first described this in terms of my dad and cars from the 1940s, when people could fix their own cars with reconditioned parts. While it is true that cars are far more reliable than they used to be, they are difficult and expensive to repair now. As they turn into electric rolling computers, they should come apart like computers, where you can "plug and play," upgrade, and change components.

Inequity: a society based on owning a vehicle is a major source of inequity; if you don't have a car, it can be impossible to get a job. We need massive investment in transit and micro-mobility alternatives.

The E-Cargo Bike

A transportation revolution is happening right now, and it's not the electric car, which just changes what powers the car but doesn't change the fundamental function of the vehicle.

The revolution happening under our noses is the rise of the e-bike, and especially the electric cargo bike. It is a new kind of vehicle that may well change the nature of our cities and suburbs. It has come out of nowhere and it has barely started, but it's a tale of the power of sufficiency to change the world.

There is nothing new about the cargo bike; a hundred years ago, they were common everywhere, and primarily used by businesses for deliveries. Although closely identified with Netherlands and Denmark, the bakfiets, or box bike, was invented in England when a clever bike builder put a box in front of the driver. The wheels were fixed to the box, and one steered by turning the entire box. The Danes developed the two-wheeled "long John" with space for carrying loads between the front wheel and the driver. In North America, the Schwinn Cycle Truck, a bike with a smaller front wheel under a big basket or shelf, was popular and marketed as a money-maker.

> In the more highly competitive fields of retail selling the difference between profit and loss is generally measured in terms of service. This is particularly true of grocery, drug and hardware stores, of meat markets, florists and beverage stores, where survival largely depends upon fast delivery service. The Schwinn Cycle Truck has solved this problem for progressive merchants from one end of the country to the other. The Cycle Truck is not an ordinary bicycle with a basket fastened to the handlebars; such makeshifts are a menace on the highways. The Schwinn Cycle Truck was designed especially for fast, local, neighborhood delivery service.

Cargo bikes were the purview of strong young men doing deliveries over relatively short distances in cities with flat terrain, and were put out of business after the Second World War not by the delivery van, but by the private automobile; the customer drove to the supermarket rather than the butcher and baker delivering to the home.

But cargo bikes didn't completely disappear. In 1998, mechanical engineer Niels Holme Larsen developed the three-wheeled Nihola specifically to carry two children and a load of groceries in a light and stable bike as an alternative to cars. There are now ten thousand of them in Copenhagen alone.

But cargo bikes are heavy, especially when loaded with kids and groceries. And while electric motors were first added to bikes in 1895, the batteries were heavy and the range was short. According to e-bike motor manufacturer Bosch, the invention of the lithium battery in 1991 made e-bikes practical; Bosch started building drives specifically for bikes in 2009 and is now a world leader. E-bike sales are now booming all over the world.

This section could be about the e-bike revolution, which is certainly in full swing, but I believe that we are in a transitional phase and that the e-bike will be subsumed into a more significant e-cargo bike revolution. I suspect that in a few years, most e-bikes will be a form of cargo bike. Just look at what is happening now. I have compared it to the Cambrian explosion 540 million years ago, which filled the seas with what *Nature* calls "an astonishing diversity of animals." It all happened in what seems like overnight in geological times. Similarly, it seems that bikes are evolving overnight and going in a hundred different directions. *Nature* calls it "the most significant event in Earth evolution."

The addition of the battery to the bicycle is likely the most significant event in bike evolution in the last 150 years, since bike designs settled on the Rover safety bicycle in 1885. The battery opens up design opportunities, and we are seeing an explosion of ideas. And I believe that in the end, the most successful designs will be a form of cargo bike.

I ride a Dutch Gazelle e-bike. It looks very much like a traditional Dutch utility bike, with regular bike wheels and a carrier, with a Bosch motor in the middle. It's designed to meet the European rules for pedelecs—250-watt motor, 20 kmph (12.4 mph) speed limit, no throttle. It is designed to be a bike with a boost and to play nicely in the bike lanes. But it is not a car replacement; I cannot carry very much on it. I could put a baby seat on the rear carrier, but the center of gravity is high.

Other manufacturers have realized that with a motor doing much of the work, they can design the bike differently. The increased rolling resistance of smaller, fatter wheels doesn't matter, so they can lower the center of gravity and carry more kids and more stuff. Bike design is evolving and adapting to take advantage of electric power; it is no longer for just getting a person from A to B; it is a family hauler. It's good for a grocery run. For many families, the e-cargo bike replaces the second car; for some, it has become the main family vehicle.

E-cargo bikes are expensive; a two-wheeled Tern can cost $6,000, and a three-wheeled Black Iron Horse can cost twice that. But prices are dropping fast. Lectric, a reputable manufacturer, just introduced a cargo bike for under $2,000. This is where it all gets very interesting.

According to Tesla, the Model 3 has upfront carbon emissions of 10.4 tons. According to my calculations from my book, *Living the 1.5 Degree Lifestyle*, my Gazelle e-bike has upfront carbon emissions of 82 kilograms (180.7 pounds) or 1/126 of the Tesla. For cargo bikes versus electric pickup trucks, the ratio might be similar.

But the differences don't stop there. The smaller batteries use 1/300 of the lithium of an F-150 Lightning. I get three e-bikes and one big cargo bike in our garage, with space left for my rowing machine and my wife's elliptical.

There are a couple of days per year when I cannot ride my e-bike, and they are usually the days that the police tell drivers to stay off the roads. There are a couple of days when I am riding in the rain, but I have good rain gear left over from my hiking days. There is no reason that an e-bike or e-cargo bike couldn't replace most trips people take in cars.

More importantly, 74 percent of North Americans live in the suburbs, where the roads are wide and the distances longer. There's lots of room to build decent separated bike infrastructure, and e-cargo bikes eat up the miles. This is where the e-cargo bike revolution will happen. An English study, "E-bikes and Their Capacity to Reduce Car CO_2 Emissions" found that their biggest impact would be in the suburbs; people who live in cities have shorter distances and more options, such as transit or walking.[44] It's the car-dependent suburbs that have greater untapped potential for e-bikes. Their study did a statistical analysis to determine what percentage of the population was fit to ride an e-bike while carrying 15 kilograms (33 pounds), equivalent to groceries or a small child, and definitely e-cargo bike territory for a 20-kilometer (12.4-mile) round trip. They concluded that "Mass uptake of e-bikes could make a significant early contribution to transport carbon reduction, particularly in areas where conventional walking and cycling do not fit journey patterns and bus provision is relatively expensive, inflexible and, certainly in the UK, has diminished over recent decades."

In my own family, the cargo bike has been transformative. My son-in-law is an urban creature who never learned to drive a car; I was shocked to find that he had also never been on a bicycle, being used to walking and transit. When their daughter Edie came along, he became dependent on his wife, my daughter Emma, whenever they had to go anywhere. He also had a pop-up business making calzones and needed Emma to drive him to every event, which was much less convenient now that there was also a baby in tow.

Then they bought a cargo bike, a Danish three-wheeler called a Black Iron Horse. It is an eccentric design with rear-wheel steering like a forklift truck. Like the truck, it is incredibly maneuverable at low speeds and can turn on a dime. Also like the truck, it is less stable at higher speeds, which, fortunately, Neil is aware of and takes it relatively slow. It costs as much as their used car, but gets far more use. I suspect if they had the cargo bike, they might not even have bought the car; they rarely use it now.

In Denmark, most cargo bikes are three-wheelers; in Netherlands, most are two-wheelers, stretched bikes with a cargo bay in the middle. I am told that it is because Netherlands had better bike infrastructure, and the Danes needed a more stable bike. That doesn't sound convincing, but they continue: "The advantage of a two-wheeled cargo bike is manifold; one less wheel on the ground means you can do longer distances with far less effort, and of course, it feels more like a regular bike." This is true, although the electric motor levels the playing field.

The micromobility expert Horace Dediu argued that "electric, connected bikes will arrive en masse before autonomous, electric cars. Riders will barely have to pedal as they whiz down streets once congested with cars." He concluded: "E-bikes will eat cars." I go further and suggest that e-cargo bikes will eat pickup trucks.

If governments took a fraction of the money they are investing in electric car subsidies, changing infrastructure and likely road rebuilds to deal with all the damage from the extra weight, they could build out fabulous bike networks and give everyone an e-bike.

Insight: An e-bike or e-cargo bike can change your life, and for many people, replace or seriously reduce the use of a car.

CHAPTER 4

EVERYTHING CONNECTS

After posting this chart from the Environmental Protection Agency, the question was raised, "If you had unlimited time, staff, and resources, which of the five major GHG-polluting sectors of the US economy would you spend your time working to decarbonize? And why? You can only pick one." This bothered me because I keep quoting transportation planner Jarrett Walker: "Land use and transportation are the same thing described in different languages." They are not separate pieces of the pie. This is why I call them "built environment emissions."

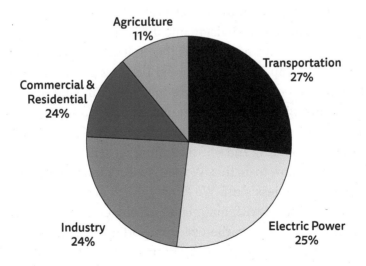

Total US Greenhouse Gas Emissions by Economic Sector in 2020.
Credit: Environmental Protection Agency.

As I noted in *Living the 1.5 Degree Lifestyle*, buildings and cars together are "a single entity or system that has evolved and expanded over the years through the changes in the form of energy available, and in particular the ever-increasing availability and reduction in the cost of fossil fuels." Basically, we are driving between buildings.

Then, there is electric power. If you look at the Lawrence Livermore Sankey chart in the chapter on electricity, you will see that 72 percent of all electricity goes into our residential and commercial buildings, with the balance going into industry, which pretty much makes them built environment emissions.

And what is industry, 24 percent of the emissions pie, doing? It's making cement for buildings and car infrastructure, running refineries to make transport fuels (a surprising 18 percent of emissions from industry), and making steel. And what is steel being used for? Sixty-six percent of it goes to construction and the auto industry.

Total it all up and we have approximately 75 percent of carbon emissions in the United States coming from land use and transportation, from buildings and cars.

There is no pie, no division into different components; it is all one thing, and we have to deal with it all as one thing. As the British group Built Environment Declares put it, "If we are to reduce and eventually reverse the environmental damage we are causing, we will need to re-imagine our buildings, cities and infrastructures as indivisible components of a larger, constantly regenerating and self-sustaining system."

We know how to do this. We have to work the demand side; we have to use less stuff. We know how to decarbonize land use and buildings with materiality, efficiency, and electricity. We know how to decarbonize transportation with electricity and frugality. But as with everything, it all comes back to sufficiency.

But if we don't keep making and selling so much stuff, what do we do? What does this world look like?

A Prosperous Ascent

In Chapter 1, we looked at Samuel Alexander's "The Sufficiency Economy: Envisioning a Prosperous Descent" and his vision of what a world based on sufficiency would look like. Much has changed since 2012 when he wrote that, and I am not convinced about his choice of categories, just as in the previous chapter, I was not convinced about the divisions of the carbon emissions pie. I also have trouble calling it a descent; I am reminded of the Richard Fariña novel *Been Down So Long It Looks Like Up to Me.* This is an ascent to a greener, better world, not a descent. However, I cannot argue with his basic premise:

> If the global population is to live safely within the sustainable carrying capacity of the planet, we must be prepared—especially those of us in the developed regions of the world—to reimagine the good life by embracing "simpler ways" of living based on notions of moderation, frugality, appropriate technology, and suffi-ciency. These notions are rarely discussed in mainstream environmental literature, and they are unspeakable by our politicians.[1]

As noted previously, work, transportation, and housing are all closely intertwined and interrelated, but for convenience and continuity reasons, let's use Alexander's categories.

Water

In 1989, UN Secretary-General Boutros Boutros-Ghali said, "The next war will be over water." While this did not turn out to be true (it was over oil), water and how it is shared among nations and states will be increasingly contentious. In the southwestern United States, irrigation and agriculture account for over 60 percent of water withdrawals, often to grow water-intensive crops such as strawberries, lettuce, and almonds. This will have to change.

In *Living the 1.5 Degree Lifestyle*, I suggested that we have to stop using fresh water to wash away poop and pee, and should go "off-pipe" with composting toilets. I described the Bullitt Center's bathrooms in Seattle as "the sweetest smelling ones I have ever used. Composting toilet systems are all under negative pressure, sucking air down through the toilet, so they never smell like bathrooms." I thought that this was the future of the toilet.

The Bullitt Center celebrated their toilets for not only saving water, but reducing carbon emissions, noting that "in California, water-related energy use consumes 19 percent of the state's electricity, 30 percent of its natural gas and 88 billion gallons of diesel fuel every year." That's a lot of upfront carbon.

But in 2021, they ripped all the composting toilets out. Some of the reasons were technical; it was hard to schedule compost management because men make more poop than women. (Solution: unisex toilets.) Managing one building was difficult. But the biggest problems were based on user experience: "There were far greater maintenance issues with the foam flush system than it had anticipated. Fully one-half of all the building engineer's time on-site was spent dealing with problems in the composters or the toilets, and the work was often quite unpleasant."

In less polite terms, the foam toilets didn't quite do the job they were supposed to, and the bowls were often soiled. This is a cultural rather than a technical problem; North Americans are used to big bowls with lots of water. In Europe, every toilet has a brush beside it, and people are expected to use it. On Quora, Europeans explained why they use the brush, even in restaurants and hotels:

> It's not about embarrassment, it's about being responsible. It's the housekeeper's job to clean your room, but the pieces of your poop in the toilet is too personal and would definitely gross out the housekeeper. Me and many other people think leaving the toilet dirty like that is rude for this very reason.

This is something North Americans are going to have to get used to. It's not hard, and it's not disgusting, it is just good manners. It will be even more of a challenge if we get urine-separating toilets, which might be necessary if we are going to try to recover the phosphorus from pee and the nitrogenous fertilizer from poop, turning our waste into a resource instead of a problem.

The key strategies for dealing with water are **frugality**— designing systems to use less or no water, and to a degree, **circularity**—reusing water and our waste instead of flushing it away.

Food

Many have pointed out for years that we are not eating food, we are eating fossil fuels. This became obvious to everyone during the winter of 2022–2023 during the Russo-Ukranian war, when many vegetables disappeared from the store shelves in February as the price of natural gas went through the roof; it turns out they were often grown under glass with LED lights and heating from natural gas and shipped long distances in diesel-powered refrigerated trucks. They were fed fertilizers made from natural gas. They were packaged in plastic made from oil and gas.

We had become used to eating lettuce and tomatoes in February instead of root vegetables like our grandparents ate in winter. But by acknowledging the **intermittency** of food and eating seasonal, local, or preserved reduces the upfront carbon emissions significantly.

There is also the question of what we eat; here we are going to need some **flexibility**. According to the *Economist*, we are in the midst of a global rice crisis.[2] "Rice feeds more than half the world—but also fuels diabetes and climate change."

Rice's contribution to global warming represents an underappreciated feedback loop. Irrigating paddy fields starves the underlying soil of oxygen. This encourages methane-emitting bacteria to flourish. Consequently, rice production is responsible for 12% of total methane emissions—and 1.5% of total greenhouse-gas emissions, comparable to aviation. Vietnam's paddy fields produce more carbon equivalent than the country's transportation.

Then there is the question of red meat. The UN Food and Agriculture Organization (FAO) estimates that raising livestock is responsible for 7 percent of global greenhouse gas emissions.[3] As noted in the section on hamburgers, this number is controversial, but Hannah Ritchie of Our World in Data is blunt: "Ignoring food emissions is simply not an option if we want to get close to our international climate targets. Even if we stopped burning fossil fuels tomorrow—an impossibility—we would still go well beyond our 1.5°C target, and nearly miss our 2°C one."

The biggest question here is whether the world can feed itself without fertilizers and run our food system on compost and poop. According to Hannah Ritchie, "It's estimated that nitrogen fertilizer now supports approximately half of the global population."[4] Two percent of the world's CO_2 emissions come from making the ammonia in that fertilizer, emitted when separating hydrogen from carbon in natural gas. Vaclav Smil, writing in *Energy and Civilization: A History*, notes the importance of this:

No other energy use offers such a payback as higher crops yields resulting from the use of synthetic nitrogen: by spending roughly 1% of global energy, it is now possible to supply about half of the nutrient used annually by the world's crops. Because about three quarters of all nitrogen in food proteins come from arable land, almost 40% of the current global food supply depends on the Haber-Bosch ammonia synthesis process. Stated in reverse, without Haber-Bosch synthesis the global population enjoying today's diets would have to be almost 40% smaller.

Smil doesn't acknowledge that about 40 percent of the food we make is wasted; dealing with that would go a long way to reducing our need for fertilizer. We also are using a lot of fertilizer to grow food for animals and even for feeding ethanol and biofuel to cars, and soon, airplanes.

Smil wrote this in 2017, but much has changed in the short time since. Production of "green" hydrogen made from the electrolysis of water using renewable electricity can replace the "gray" hydrogen made from natural gas, giving us carbon-free fertilizer. Given that half the planet relies on it, this is likely a more realistic scenario than relying on compost and poop.

So, the key strategies for dealing with food are **intermittency**—learning to eat with the seasons, **flexibility**—adapting our diets to foods with lower upfront carbon emissions, and **satiety**—eating less and wasting less.

Clothing

Being a white male architect of a certain age, clothing was not a consideration when I wrote *Living the 1.5 Degree Lifestyle*; I just wear the same old architect's uniform of black on black.

Yet according to C40 Cities, clothing and textiles are responsible for 4 percent of greenhouse gas emissions, bigger than aviation and electronics. Much of this is "fast fashion" made from polyester, which is made from fossil fuels and dissolving into dangerous microplastics. But natural materials have their own issues; the water intensity of cotton and the methane from sheep. Linen made from flax might be a better choice, or hemp. Tencel, a cellulose rayon made without harmful chemicals, might be our choice of near-synthetics. But the best approach is once again **circularity**, or repair and reuse. And **satiety** because maybe we don't need so much clothing.

This is already happening. My daughter has a dedicated sewing room and is making almost all of her and her daughters' clothing. As I write this, she is turning a worn tablecloth of my late mother's into a pouffe. Secondhand clothing has become so popular that we are getting the gentrification of thrifting, where people without much money—the usual customers of thrift stores—are being priced out of the market.[5] But the future may well be very much like those posters from the Second World War, the British "Make Do and Mend," and the American "Use It Up, Wear It Out, Make It Do, or Do Without." Or my favorite: "Remember Pearl Harbor/Purl Harder!" Sarah Sundin writes: "Mending was more than economical, it was a patriotic duty, and a fad for patched clothing emerged. Home sewers often pieced together garments from remnants, mixing and matching colors and patterns. Creative women cut down old garments to reuse the cloth and remade old clothing into wartime fashions."[6] Alexander thought fashion would just go away, but noted dismissively "that clothing's purpose has evolved to become primarily about expressing one's identity or social status." I suspect this will remain true, just as it did during the war. And architects will still be wearing black.

Transportation

Remember the self-driving car? By now, they were supposed to be everywhere. They would be electric and shared, and since they were always on the move, 95 percent of parking spaces would disappear. Now, all the hype is about electric cars. But if we design our communities properly, we won't need cars. With the rising costs of vehicles, insurance, and financing, fewer people can afford cars. Most importantly, with the upfront carbon emissions of manufacturing cars and car infrastructure, we have no carbon budget headroom for cars. We also now have alternatives; as Horace Dediu noted, "Electric, connected bikes will arrive en masse before autonomous, electric cars. E-bikes will eat cars." But as noted many times, if we are going to have an e-bike revolution, we also need safe places to ride and secure places to park.

Perhaps the most intractable transportation problem is aviation; we are promised everything from hydrogen to sustainable aviation fuels, but they are unrealistic. What we can do is build out high-speed rail networks as they did in China, where they covered the country in two decades. This could eliminate most short-haul or internal flights. According to the International Civil Aviation Organization, only 30 percent of flights are over water; eliminating the other 70 percent would go a long way to ameliorating the problem of aviation. Unfortunately, building out a high-speed rail network in North America is probably a fantasy; the longest line in China is 2,298 kilometers (1,427.9 miles), and it is 4,411 km (2,740.8 miles) from Toronto to Vancouver. But at 300 kmph (186.4 mph), the speed I have travelled in China, I could get there in fifteen hours.

Ultimately, the only realistic solution to the problem of flying is satiety, or enoughness; we will just have to do less of it. Most of the world doesn't fly at all; for the rich who do, they should travel as they did a hundred years ago with grand tours, staying much longer but traveling less often.

Housing

At the time of writing, 27 percent of San Francisco's office space is vacant, a total of 21 million square feet. Calgary is at 32 percent and Toronto at 18 percent and rising, as the way we work changes and as jobs vanish. Much of this is in Class B and C buildings, which need significant upgrades in mechanical systems before anyone will occupy them after the pandemic. All these cities have hundreds, if not thousands, of people needing housing.

Refurbishing, renovating, and repurposing these buildings may well be the major construction challenge of the next few years. It's not always easy; office floor plates are often much deeper than in residential buildings. But there are creative approaches; Perkins + Will architects are taking suburban office buildings and knocking out the middle, turning them into courtyards. Also, since more people will be working from home, the core areas of buildings might turn into workshops, storage areas, kitchens, or office uses with a window way in the distance.

All of this will be necessary to bring enough people back to our city cores to support the main streets, the stores, and the businesses people need to survive because they can't drive to Walmart and the big box stores; they likely won't own cars.

There has been much outrage recently about what Professor Carlos Moreno called a fifteen-minute city, where you can get everything you need within a short walk. Work, doctors, education, entertainment, and shops are all available within walking distance. But it is really nothing new; it is also a description of just about every city neighborhood or town designed before 1920, designed and built as streetcar suburbs. You got most of what you needed within fifteen minutes, and could hop on the streetcar if you couldn't find it. The density was high enough so that most people could walk to the streetcar or the Main Street in a reasonable time; nobody needed a car. When I wrote *Living the 1.5 Degree Lifestyle*, I found that living in a streetcar suburb was the single biggest determinant of my carbon footprint. I never had to get into a car; I could buy almost everything I needed within walking distance from a house that was easily subdivided into a multiple family residence.

Meanwhile, across the street, a house that was duplexed many years ago is being expensively enlarged and turned back into a single-family dwelling. This is an unfortunate trend; whole swathes of Toronto where big houses were turned into apartments after the Second World War have been converted back, dropping the population density significantly. Rental buildings are being demolished to build bigger condo buildings. Between the two trends, affordable housing is disappearing.

We need instead to follow the European model, with strong rent controls, higher densities, with small apartment buildings built around single open stairs. As noted in the block of flats section, they have the lowest carbon footprint of any built form.

Where we do build single-family housing, we have to build

it to a higher efficiency standard, preferably Passivhaus, from materials with low upfront carbon. But we also need to take sufficiency into account. Professor Kevin Anderson of Manchester University asks, "Why are we building homes that are 200 to 400 m² (2,143 to 4,306 square feet)? Cut this to a maximum of 100 to 150 m² (1,076 to 1,614 square feet), still large homes, but with much less resource and material use—and of course less land!"[7] For much of the world, 1,614 square feet is palatial; in North America, not so much. But it is sufficient.

It is also likely that we will see the revitalization of our small towns, with people living closer to the sources of food, water, and wind.

Work and Production

In *Living the 1.5 Degree Lifestyle*, written during the pandemic, I noted that the third Industrial Revolution wrought by the computer is finally kicking in to dematerialize the workplace. More and more people were working from home, and I predicted that they would not be going back to the office. I was only half right; there was big pushback from management for "hybrid work," with a couple of days a week in the office and a couple at home. From a carbon emissions point of view, it is probably the worst of both worlds, requiring the maintenance of two office workspaces instead of one, and the continuing operation of the commuting infrastructure, albeit at lower intensities. This is leading to the collapse of public transit systems, and I believe the coming crash in commercial real estate.

But there is also going to be an inevitable shakeout of jobs as we deal with the problem of our built environment emissions; as I noted earlier, degrowth is "the inevitable result of sufficiency, of sobriety, of demand side mitigation, of using less stuff because the world's economy is essentially about making stuff, whether it is green or not."

The US Bureau of Labor Statistics made some interesting predictions in 2021 about where the job market is going in the next ten years, and saw dramatic growth in jobs relating to community and social services, healthcare support, food service, and personal care and service occupations.[8] This is likely the result of all of us aging baby boomers needing support. They see single-digit increases in management and declines in sales, office administration, and production. But there are other factors in play here.

The head of the Ford Motor Company says that as many as 40 percent of the workforce may be superfluous thanks to electrification, as the cars are so much simpler; more might be laid off if people are not driving to work and don't need as many cars. The US Census Bureau notes that 57.2 percent of households have two or more cars in the driveway; imagine if they didn't need that anymore and cut back to one. For that matter, 9.1 percent of Americans do not have a car at all; imagine if that increased to the Dutch level of car-free households at 33 percent.

But we have to go much further than that. We previously posited that about 75 percent of our emissions are built environment emissions; to deal with them, we must consider a world where we are not making cars. We are not building highways and parking garages for cars. We are not building concrete and steel towers. A lot of basic construction, production, transportation, and material-moving jobs will disappear, although maintenance and repair occupations will increase.

We know that the key to reducing our upfront carbon emissions is to use less stuff. To make less stuff. To reuse stuff. The most important jobs of the future might well be repairing what we have. It might be a world my dad would recognize from his teenage days chasing scrap, my daughter would recognize from her time sewing, my other daughter making bread and selling cheese, my wife from her canning and preserving, and me renovating and retrofitting buildings, with much of our spare time spent working in the garden. I suspect that we will all be busy.

CONCLUSION

When you look at the world through the lens of upfront carbon, everything changes. The idea that we can build out a new world running on renewable energy in time to prevent catastrophic climate change seems less plausible; there is so much to do in so little time. There is so much stuff that would have to be supplied, and it all comes with a hefty burp of upfront carbon. Instead, we have to work the demand side and use less stuff. The IPCC said:

> Demand-side mitigation encompasses changes in infrastructure use, end-use technology adoption, and socio-cultural and behavioural change. Demand-side measures and new ways of end-use service provision can reduce global GHG emissions in end-use sectors by 40–70% by 2050.

We can't just pretend that we can have it all, the "same-sized homes. Same-sized cars. Same levels of comfort. Just electric." That is just too much stuff. Vaclav Smil said:

> In 2021 there were some 1.4 billion motor vehicles on the road, of which no more than 1 percent were electric. Even if the global road fleet were to stop growing, decarbonizing 50 percent of it by 2030 would require that we manufacture about 600 million new electric passenger vehicles in nine years—that is about 66 million a year, more than the total global production of all cars in 2019.

In addition, the electricity to run those cars would have to come from zero-carbon sources. What are the chances of that?

And that's not even starting with the 18 gigatons of upfront carbon from making 1.4 billion cars. We learned from Smil about the power concentrated in fossil fuels, with which we are so profligate. But without the time and the materials that we need to replace them, the only option is to make and use less stuff, and to dig less stuff. Geologist Simon Michaux said:

> The logistical challenges to replace fossil fuels are enormous. It may be so much simpler to reduce demand for energy and raw materials in general. This will require a restructuring of society and its expectations, resulting in a new social contract. Is it time to restructure society and the industrial ecosystem to consume less?[1]

Us consuming less means someone else is extracting less, producing less, building less, and selling less. It will mean a different economy, where we are making less stuff. But it could be a better life for everyone. J. B. MacKinnon said:

> The evidence suggests that life in a lower-consuming society really can be better, with less stress, less work or more meaningful work, and more time for the people and things that matter most. The objects that surround us can be well made or beautiful or both, and stay with us long enough to become vessels for our memories and stories. Perhaps best of all, we can savour the experience of watching our exhausted planet surge back to life: more clear water, more blue skies, more forests, more nightingales, more whales.

Professor Anderson has similar thoughts:

> The majority of people will be better off in virtually all aspects of their lives. Not only the elimination of fuel

poverty—at last—but improved and warmer homes, reduced bills and much better indoor and outdoor air quality—leading to healthier children more able to participate fully in school. Clean, efficient and reliable public transport—for all citizens—less noise, more usable urban space for parks, cafés, playing fields and the many other facilities that make a thriving community.

As noted in the previous chapter, we will be leading different lives. Samuel Alexander said:

> This would be a way of life based on modest material and energy needs but nevertheless rich in other dimensions— a life of frugal abundance. It is about creating an economy based on sufficiency, knowing how much is enough to live well, and discovering that enough is plenty.

We can power everything with electricity. We can reuse and repurpose with circularity. We can take just what we need with satiety and frugality. We can make things right with materiality and simplicity. But in the end, it all comes down to sufficiency. Enough already.

CHEAT SHEET: A SHORT GUIDE TO SUFFICIENCY

We started this book by noting that "when you look through the lens of upfront carbon, everything changes." One of the main things you see is that we need to use less of everything.

"Less is more," the phrase popularized by architect Mies van der Rohe comes to mind, but he didn't say it first; that was poet Robert Browning in his 1855 poem "Andrea del Sarto." And Mies didn't practice what he preached; his simple, minimalist details were often expensive and impractical. Bucky Fuller defined ephemerality as "doing more with less." Degrowth advocate Jason Hickel titled his book *Less Is More: How Degrowth Will Save the World* and writes that "degrowth begins as a process of taking less." Architect Kelly Alvarez Doran rejects Mies and gets back to basics: "less is less." And that's really what this is all about: using less stuff. *Less is less.*

So, let's go back to those prosperous ascent categories and drill down to the basics:

Food

Acknowledge the intermittency of food production and eat a seasonal diet. Reduce the inefficiencies of the food chain by eating as local as possible, and as simply as possible, with few heavily processed foods. Drastically limit red meat and dairy. Reduce waste and monitor portion sizes.

Transportation

The mode with the lowest upfront carbon is walking, and we should all do more of it. We have to electrify our transport, but also recognize that electric cars won't save us; instead, we need fewer cars and must promote alternatives. I continue to believe that we are at the start of an e-bike revolution, but for those who drive, the better answer might be to figure out how to drive a lot less rather than buying an electric car. The same can be said for long-distance travel: There is no point in waiting for electric or hydrogen planes; the only realistic solution is to fly as little as possible. Take ground-based alternatives where you can and consolidate your trips to get as much as you can out of every flight you take.

Housing

Find a place to live where you don't need to drive everywhere. Don't rent or buy any more space than you need; think about enoughness. Small multifamily buildings have the lowest upfront carbon emissions and are usually found in walkable neighborhoods. Maximize universality, where you can stay there as long as possible; resiliency, where you can survive when the power goes out; and flexibility, where you can use space in multiple ways.

Stuff

From clothing to computers, so many of our properties of sustainability come into play. Frugality is important; buy quality that will last, simple designs that are robust and that can be serviced. Better still, share instead of buy.

Materiality

What stuff is made of matters too. Natural materials and fabrics have lower upfront carbon emissions than mined or manufactured materials.

How stuff is powered matters; we want everything to run on low-carbon electricity, but it will never be zero-carbon; there is the upfront cost of building the source and getting it to you. This is why we will always have to use as little of it as possible.

It's like the book cover says: A life of "just enough" offers a way out of the climate crisis.

Notes

Chapter 1: The Lens of Upfront Carbon

1. "Embodied energy," *Wikipedia*.
2. Brown, Martin, "A Carbon Hierarchy for (Net) Zero Carbon Construction," *Fairsnape*, January 7, 2020.
3. EPA, *Greenhouse Gas Emissions from a Typical Passenger Vehicle*, 2023, epa.gov.
4. EPA, *Equivalent Emissions*, 2023, epa.gov.
5. Beacon, Geoff, "Carbon Emissions in the Lifetimes of Cars," Brussels Blog, March 11, 2019.
6. Brusseau, M. L., "Sustainable Development and Other Solutions to Pollution and Global Change," *Environmental and Pollution Science*, 2019, pp. 585–603, sciencedirect.com.
7. www.withouthotair.com.
8. Natali, Paolo, Greene, Suzanne, and Toledano, Perrine, "How Much CO_2 Is Embedded in a Product?," RMI, August 5, 2019.
9. Carter, Jimmy, April 18, 1977, "A Moral Equivalent of War," *Wikisource*, April 25, 2011.
10. Samarajiva, Indrajit, "What the 1970s Oil Shock Can Tell Us About Today," indi.ca, August 1, 2022.
11. "Oil Dependence and U.S. Foreign Policy," Council on Foreign Relations, 2023.
12. Greenberg, John, "Ronald Regan's Son Says His Father Got the Saudis to Pump More Oil to Undercut USSR," *Politifact*, March 6, 2014.
13. Cleveland, Cutler J. (ed.), *Encyclopedia of Energy*, 2004, sciencedirect.com.
14. Alberta Energy and Minerals, "Oil Sands Facts and Statistics," alberta.ca.
15. Glaser, M.B., "1982 Memo to Exxon management about CO_2 Greenhouse Effect," climatefiles.com, 2023.
16. Weart, Spencer, "The Discovery of Global Warming," aip.org, May, 2023.
17. Ibid.
18. Roston, Eric, "The World Is Moving Toward Net Zero Because of a Single Sentence," *Bloomberg Green*, February 8, 2021, bloomberg.com.
19. Osaka, Shannon, "The World's Most Ambitious Climate Goal Is Essentially Out of Reach," *Grist*, April 8, 2022.
20. Westervelt, Amy, "IPCC: We Can Tackle Climate Change if Big Oil Gets Out of the Way," *The Guardian*, April 5, 2022.
21. Nuccitelli, Dana, "The 5 Stages of Climate Denial Are on Display Ahead of the IPCC Report," *The Guardian*, September 16, 2013.
22. Mann, Michael, *The New Climate War*, Public Affairs, 2022.
23. Ritchie, Hannah, "Climate Deniers and Doomers Are More Alike than They'd Like to Think," hannaritchie.substack.com, April 16, 2023.
24. Nuttall, Philippa, "Don't Listen to the Climate Doomists," *The Newstatesman*, 2023.

25. Anderson, Kevin, "A True Paradise: Where We Are Heading," *Transformative Urban Coalitions*, 2023.
26. "The 2030 Challenge for Embodied Carbon," architecture2030.org, 2023.
27. University of Washington, "Embodied Carbon 101," carbonleadershipforum.org, 2023.
28. "What's In A Footprint?," allbirds.com, 2023.
29. Roberts, David, "How to Drive Fossil Fuels Out of the US Economy, Quickly," *Vox*, August 6, 2020.
30. Grant, Nick, "Sustainable Building," *Elemental Solutions* (blog), *Sustainable Building*, 2012.
31. Michaux, Simon, "Assessment to Phase Out Fossil Fuels."
32. Alexander, Samuel, "Sufficiency Economy," 2017.
33. Sidler, Oliver, "Sobriety and Energy Efficiency: Two Complementary Approaches," *Up To Us*.
34. Ayres, Robert, "What Is Exernomics?" *Exernomics: On Energy, Economy and Growth*, October 15, 2014.
35. Kallis, Giorgos, "Radical Dematerialization and Degrowth," *Royal Society Publishing*, May 1, 2017.
36. Cox, Stan, "Needed: Either Degrowth or Two Earths," *Countercurrents*, June 2023.
37. Gore, Tim, "Confronting Carbon Inequality," Oxfam, September 21, 2020.
38. Creutzig, Felix, et al., "Demand-Side Solutions to Climate Change Mitigation Consistent with High Levels of Well-Being," *Nature*, November 21, 2021.
39. Jackson, Robert B., et al., "Human Well-Being and Per Capita Energy Use," *Ecosphere*, April 12, 2022.
40. Baltruszewicz, Marta, et al., "Social Outcomes of Energy Use in the United Kingdom: Household Energy Footprints and Their Links to Well-Being," *Ecological Economics, Vol. 205*, March 2023, Sciencedirect.com.
41. Alexander, Samuel, "A Critique of Techno-Optimism," 2017.
42. Saheb, Yamina, "COP26: Sufficiency Should Be First," *Buildings and Cities*, Ocober 10, 2021.

Chapter 2: Strategies for Sufficiency

1. "Refuse Unnecessary New Construction," *Strategies/Actions*, Circular Buildings Toolkit.

2

3. Morfeldt, Johannes, et al., "The Impact of Climate Targets on Future Steel Production," *ResearchGate*, April 2014.
4. Wu, Malan, et al., "Pedal to the Metal: Iron and Steel's US$1.4 Trillion Shot at Decarbonisation," *Wood Mackenzie*, September 2022.
5. "Aluminum," *The Essential Chemical Industry Online*, September 24, 2016.
6. "Infinitely Recyclable," *The Aluminum Association*, 2021.
7. "Apple Buys First Carbon-Free Aluminium from Alcoa-Rio Tinto Joint," *Aluminum Insider*, December 6, 2019.
8. Cichon, Steve, "Everything from this 1991 Radio Shack Ad You Can Now Do with Your Phone," *Huffpost*, January 16, 2014.
9. Peterat, Linda, "Cook Stove Revolution of the 1800s," bcfoodhistory, July 8, 2019.

10. Nguyen, Mai, "Innovation in EVs Seen Denting Copper Demand Growth Potential," *Reuters*, July 9, 2023.
11. U.S. Geological Survey, 2020, *Mineral Commodity Summaries 2020*.
12. "Copper Facts: Copper in the Home," copper.org.
13. Michaux, Simon, "GTK Reports," simonmichaux.com, 2018.
14. Henze, Veronica, "Lithium-ion Battery Pack Prices Rise for First Time to an Average of $151/kWh," *BloombergNEF*, December 6, 2022.
15. Pinsker, Joe, "Frugality Isn't What it Used to Be," *The Atlantic*, October 22, 2016.
16. Kumar, Nirmalya, et al., "Frugal Engineering: An Emerging Innovation Paradigm," *Ivey Business Journal*, March/April 2012.
17. Ibid.
18. www.innofrugal.org
19. www.frugalengineering.in
20. Wharton School of the University of Pennsylvania, "How to Avoid the Pitfalls of Innovation in Emerging Markets," *Knowledge at Wharton*, October 20, 2016.
21. "The Futurist: We Predict the iPhone Will Bomb," *TechCrunch*, June 7, 2007.
22. Eliason, Mike, "In Praise of Dumb Boxes," *Medium*, August 13, 2018.
23. Antonelli, Lenny, "Seeing the Wood for the Trees—Placing Ecology at the Heart of Construction," *Passivehouseplus*, October 26, 2021.
24. Kindy, David, "How the Volkswagen Bus Became a Symbol of Counterculture," *Smithsonian Magazine*, March 6, 2020.
25. Kronenburg, Robert, "Adaptable Architecture—Flexible Dwelling," *Academia*, 2005.
26. Kumar, Mohi, "From Gunpowder to Teeth Whitener: The Science Behind Historic Uses of Urine," *Smithsonian Magazine*, August 20, 2013.
27. Bockmann, Michelle Wiese, "Shipping Emissions Rise 4.9% in 2021," *Lloydslist*, Maritime Intelligence, January 24, 2022.
28. Korhonen, Jouni, "Circular Economy: The Concept and Its Limitations," *Ecological Economics*, Vol. 145, January 2018, Sciencedirect.com.
29. Dix, T. Keith, "'The Remains of My Books': Cicero's Library at Antium," *Camws.org*.
30. Baltruszewicz, Marta, et al., "Social Outcomes of Energy Use in the United Kingdom: Household Energy Footprints and Their Links to Well-Being," *Ecological Economics*, Vol. 205, March 2023, Sciencedirect.com.
31. Gabbatiss, Josh, "Richest People in UK 'Use more Energy Flying' Than Poorest Do Overall," *Carbon Brief*, December 4, 2022.
32. Stiglitz, Joseph, "The Climate Crisis Is Like a World War So Let's Talk Rationing," *The Globe and Mail*, December 14, 2019.
33. Griffith, Saul and Calisch, Sam, "No Place Like Home: Fighting Climate Change (and Saving Money) by Electrifying America's Households," *Household Savings Report*, October 2020, rewiringamerica.org.
34. Barnard, Michael, "With Heat from Heat Pumps, US Energy Requirements Could Plummet by 50%," *Clean Power*, March 14, 2023, cleanTechnica.com.
35. Buonocore, Jonathan J., "Inefficient Building Electrification Will Require Massive Buildout of Renewable Energy and Seasonal Energy Storage," *Scientific Reports*, July 13, 2022, Nature.com.
36. Brown, M. J., "Too Cheap to Meter?," *Web Archive*, December 14, 2016.
37. Helman, Christopher, "How Green Is Wind Power, Really? A New Report Tallies up the Carbon Cost of Renewables," *Forbes*, April 28, 2021.

38. Fawkes, Steven, "Fawkes's Laws of Energy Efficiency," *Only Eleven Percent*, December 21, 2015.

39. Perez, Marc, et al., "Overbuilding & Curtailment: The Cost-Effective Enablers of Firm PV Generation," *Solar Energy*, Vol. 180, March 1, 2019.

40. Fares, Robert, "Renewable Energy Intermittency Explained: Challenges, Solutions, and Opportunities," *Scientific American*, March 11, 2015.

41. Roche, David, "Using Electric Water Heaters to Store Renewable Energy Could Do the work of 2 Million Home Batteries and Save Us Billions," *The Conversation*, June 4, 2023.

42. Liebreich, Michael, "Separating Hype from Hydrogen Part Two: The Demand Side," *BloombergNEF*, October 15, 2020.

43. Melton, Paula, "The Urgency of Embodied Carbon and What You Can Do About It," *Building Green*, Vol. 27 (9).

44. Thompson, Clive, "Electric Cars Are Weirdly Skeuomorphic," *Medium*, December 9, 2021.

45. Chancel, Lucas, et al., "Climate Inequality Report 2023, Fair Taxes for a Sustainable Future in the Global South," *World Inequality Database*, January 30, 2023.

46. Kanitkar, Tejal, et al., "Equity Assessment of Global Mitigation Pathways in the IPCC Sixth Assessment Report," *OSF Preprints*, December 8, 2022.

47. Evans, Simon, "Analysis: Which Countries Are Historically Responsible for Climate Change?," *Carbon Brief*, October 5, 2021.

48. Nandi, Jayashree, "IPCC's Climate Change Mitigation Scenario Inequitable, Says Study Ahead of COP27," *Hindustan Times*, November 6, 2022.

Chapter 3: Stuff

1. "The Miracle of Carbon," *Fermilab*, August 8, 2019.

2. Park, Michael Y., "A Brief History of the Disposable Coffee Cup," *Bon Appetit*, May 30, 2014.

3. Gannon, Devin, "The History Behind NYC's Iconic Anthora Coffee Cups," *6sqft*, April 7, 2021.

4. Littman, Julie, "90% of New Starbucks Stores Will Have Drive-Thrus," *Restaurant Dive*, May 4, 2022.

5. "Paper Coffee Cups," goodstartpackaging.com.

6. "Waste Paper Piles of Canadian Brand Names Manually Segregated from Plastic," cbc.ca, June, 2023.

7. Martinko, Katherine, "Why We Need to Start Drinking Coffee Like Italians," *Treehugger*, October 25, 2022.

8. Wagner, Kate, "Q&A: Paul Andersen and Paul Preissner on American Framing," *Metropolis*, May 21, 2021.

9. Dauksta, Dainis, "*Brettstapel*, A New Name for an Old Technique," *Fourthdoor*.

10. Durham, Steve, et al., "Mass Timber and Sustainability," *Kirksey Architecture*, April 11, 2023.

11. "Mass Timber FAQ," *Mass Timber Institute*.

12. Hawkins, Will, "Timber and Carbon Sequestration," *The Structural Engineer*, January 2021.

13. Alter, Lloyd, "Total Carbon Footprint of a North American Home Is More Than You Think," *Treehugger*, April 13, 2022.

14. Antonelli, Lenny, "Seeing the Wood for the Trees."

15. Crook, Lizzie, "Building Tall with Timber 'Does Not Make Sense' Say Experts," *Dezeen*, March 29, 2023.

16. Gauch, H. L., et al., "What Really matters in Multi-Storey Building Design? A Simultaneous Sensitivity Study of Embodied Carbon, Construction Cost, and Operational Energy," *Applied Energy*, Vol. 333, March 1, 2023, sciencedirect.com.

17. Holladay, Martin, "Reassessing Passive Solar Design Principles," *Green Building Advisor*, October 9, 2015.

18. Creutzig, Felis, et al., "Demand-Side Solutions to Climate Change Mitigation."

19. Richardson, Jo and Coley, David, "Labour's Low-Carbon 'Warm Homes for All' Could Revolutionise Social Housing—Experts," *The Conversation*, November 5, 2019.

20. Bridger, Jessica, "Don't Call It a Commune: Inside Berlin's Radical Cohousing Project," *Metropolis*, June 10, 2015.

21. Eliason, Mike, "Better Living Through Baugruppen: A New Approach to Affordable Urban Living," larchlab.com.

22. www.icea.bio.

23. Qian, Weiran, et al., "Carbon Footprint and Water Footprint Assessment of Virgin and Recycled Polyester Textiles," *Textile Research Journal*, April 9, 2021, journals.sagepub.com.

24. "Greenhouse Gas Emissions per Kilogram of Food Product," *Our World in Data*.

25. Webber, Jemima, "A Big Mac's Carbon Footprint Is Equal to Driving a Car Nearly 8 Miles, New Data Shows," *Plant Based News*, November 4, 2021.

26. "Beef Carbon Footprint & Environmental Impact," *Consumer Ecology*.

27. Mitloehner, Frank and Hudson, Darren, "No, Four Pounds of Beef Doesn't Equal the Emissions of a Transatlantic Flight," *Ghgguru Blog*, September 26, 2019.

28. Koop, Avery, "Ranked: The Most Popular Fast Food Brands in America," *Visual Capitalist*, August 31, 2022.

29. "Korech—The Hillel Sandwich," *Jewish Practice*, Chabad.org.

30. "SeeLevel HX 21st Annual Drive-Thru Study Uncovers Delays and Inaccuracy as QSRs Struggle with Labor Shortage," prnewswire.com, September 23, 2021.

31. www.oica.net

32. www.autosinnovate.org

33. "Cost of Auto Crashes & Statistics," *Rocky Mountain Insurance Information Association*.

34. "Leading Manufacturers Support Move Towards Better Emissions Measurement for the Automobile Industry," *World Business Council for Sustainable Development*, April 12, 2022.

35. "Estimated Worldwide Motor Vehicle Production from 2000 to 2022," *Transportation & Logistics*, statista.com, 2023.

36. Seo, Sarah A., "The New Public," *Yale Law Journal* 125:1616, 2016.

37. Wood, Ruth, "Why the Automotive Industry Is Pointing Headlights at Suppliers in the Race to Cut Carbon Emission," *Emitwise*, July 18, 2022.

38. Hannon, Eric, et al., "The Zero-Carbon Car: Abating Material Emissions Is Next on the Agenda," *McKinsey Sustainability*, September 18, 2020.

39. Shan, Yuli, "Peak Cement-Related CO_2 Emissions and the Changes in Drivers in China," *Journal of Industrial Ecology Vol. 23 (4)*, August 2019.

40. Laurent, Andrew Paul, "Study: Caron-Neutral Pavements Are Possible by 2050, but Rapid Policy and Industry Action Are Needed," *MIT News*, February 21, 2023.

41. Costa, Kristina, et al., "Reducing Carbon Pollution Through Infrastructure," *American Progress*, September 3, 2019.
42. www.shift.rmi.org
43. Lewis, Matthew, "To Solve Climate, We Need Electric Cars—And a lot less Driving," *substack.com*, December 21, 2022.
44. Philips, Ian, et al., "E-Bikes and their Capability to Reduce Car CO_2 Emissions," *Transport Policy*, Vol. 116, sciencedirect.com, February 2022.

Chapter 4: Everything Connects

1. Alexander, Samuel, "Introduction to 'Prosperous Descent'," *The Simplicity Collective*, May 20, 2015.
2. Akbarpur, Bassi, "The Global Rice Crisis," *The Economist*, March 28, 2023.
3. www.fao.org
4. Ritchie, Hannah, "How Many People Does Synthetic Fertilizer Feed?," *Our World in Data*, November 7, 2017.
5. Ho, Sena, "How the Gentrification of Thrifting Affects Low-Income Communities," *The Borgen Project*, September 13, 2022.
6. Sundin, Sarah, "Make It Do—Clothing in World War II," *Sarah Sundin* (blog), March 8, 2022.
7. Anderson, Kevin, "Getting Real: What Would Serious Climate Action Look Like?," sgr.org, March 19, 2023.
8. "Occupational Projections and Worker Characteristics," *Employment Projections*, U.S. Bureau of Labor Statistics, September 6, 2023.

Conclusion

1. Michaux, Simon, "Time to Discuss," It's Time to Wake Up, gtk.fi.

Index

About the Author

LLOYD ALTER is a writer, public speaker, architect, inventor, and Adjunct Professor of Sustainable Design at Toronto Metropolitan University. He has published many thousands of articles on *TreeHugger* where he was Design Editor, and on such diverse platforms as *Planet Green*, *HuffPo*, *The Guardian*, *Corporate Knights Magazine*, and *Azure Magazine*. A former builder of prefab housing and a tiny-house pioneer, Lloyd is a passionate advocate of Radical Sufficiency—the belief that we use too much space, too much land, too much food, too much fuel, and too much money, and that the key to sustainability is to simply use less. He is the author of *Living the 1.5 Degree Lifestyle*. Lloyd lives in Toronto, Ontario.

ABOUT NEW SOCIETY PUBLISHERS

New Society Publishers is an activist, solutions-oriented publisher focused on publishing books to build a more just and sustainable future. Our books offer tips, tools, and insights from leading experts in a wide range of areas.

We're proud to hold to the highest environmental and social standards of any publisher in North America. When you buy New Society books, you are part of the solution!

At New Society Publishers, we care deeply about *what* we publish—but also about *how* we do business.

- This book is printed on 100% **post-consumer recycled paper**, processed chlorine-free, with low-VOC vegetable-based inks (since 2002)
- Our corporate structure is an innovative employee shareholder agreement, so we're one-third employee-owned (since 2015)
- We've created a Statement of Ethics (2021). The intent of this Statement is to act as a framework to guide our actions and facilitate feedback for continuous improvement of our work
- We're carbon-neutral (since 2006)
- We're certified as a B Corporation (since 2016)
- We're Signatories to the UN's Sustainable Development Goals (SDG) Publishers Compact (2020–2030, the Decade of Action)

To download our full catalog, sign up for our quarterly newsletter, and to learn more about New Society Publishers, please visit newsociety.com.

ENVIRONMENTAL BENEFITS STATEMENT

New Society Publishers saved the following resources by printing the pages of this book on chlorine free paper made with 100% post-consumer waste.

TREES	WATER	ENERGY	SOLID WASTE	GREENHOUSE GASES
20	1,600	8	68	8,670
FULLY GROWN	GALLONS	MILLION BTUs	POUNDS	POUNDS

Environmental impact estimates were made using the Environmental Paper Network Paper Calculator 4.0. For more information visit www.papercalculator.org

Certified
Ⓑ
Corporation

new society
PUBLISHERS
www.newsociety.com

FSC
www.fsc.org

MIX
Paper | Supporting responsible forestry
FSC® C016245

SDG PUBLISHERS COMPACT